Professor Edzard Ernst is the world's first professor of complementary medicine. Formerly a clinical doctor, he has studied homeopathy and practised this and many other alternative treatments. He has now built a world-class reputation for successfully applying science to test the value of alternative therapies and is regularly interviewed for TV and radio.

Dr Simon Singh has a Ph.D. in particle physics and started his career as a producer on BBC science programmes such as *Horizon* and *Tomorrow's World*. He is the author of the world-wide bestsellers *Fermat's Last Theorem* and *The Code Book*. *Big Bang*, his history of cosmology, became a *New York Times* bestseller.

Dorset Libraries
Withdrawn Stock

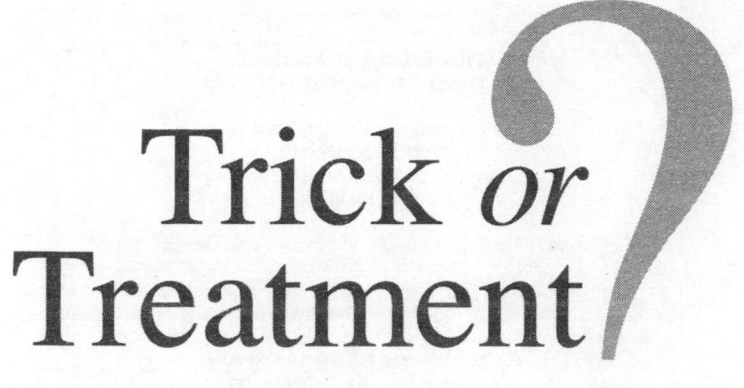

Trick *or* Treatment?

ALTERNATIVE MEDICINE ON TRIAL

Simon Singh & Edzard Ernst

Dorset Libraries Withdrawn Stock

Dorset County Library	
Askews	2009
615.5	£8.99

CORGI BOOKS

TRANSWORLD PUBLISHERS
61–63 Uxbridge Road, London W5 5SA
A Random House Group Company
www.rbooks.co.uk

TRICK OR TREATMENT
A CORGI BOOK: 9780552157629

First published in Great Britain
in 2008 by Bantam Press
a division of Transworld Publishers
Corgi edition published 2009

Copyright © Dr Simon Singh and Professor Edzard Ernst 2008

Simon Singh and Edzard Ernst have asserted their rights under the Copyright,
Designs and Patents Act 1988 to be identified as the authors of this work.

A CIP catalogue record for this book
is available from the British Library.

This book is sold subject to the condition that it shall not,
by way of trade or otherwise, be lent, resold, hired out,
or otherwise circulated without the publisher's prior
consent in any form of binding or cover other than that
in which it is published and without a similar condition,
including this condition, being imposed on the
subsequent purchaser.

Addresses for Random House Group Ltd companies outside the UK
can be found at: www.randomhouse.co.uk
The Random House Group Ltd Reg. No. 954009

The Random House Group Limited supports The Forest Stewardship Council (FSC),
the leading international forest certification organisation. All our titles that are
printed on Greenpeace approved FSC certified paper carry the FSC logo. Our paper
procurement policy can be found at www.rbooks.co.uk/environment

Typeset in 11/13.5 Times by
Falcon Oast Graphic Art Ltd.
Printed in the UK by CPI Cox & Wyman, Reading, RG1 8EX.

2 4 6 8 10 9 7 5 3 1

Every effort has been made to obtain the necessary permissions with reference to
copyright material, both illustrative and quoted. We apologize for any omissions in
this respect and will be pleased to make the appropriate acknowledgements in any
future edition.

Dedicated to

HRH The Prince of Wales

Contents

Introduction

THE CONTENTS OF THIS BOOK ARE GUIDED ENTIRELY BY A SINGLE pithy sentence, written over 2,000 years ago by Hippocrates of Cos. Recognized as the father of medicine, he stated:

> *'There are, in fact, two things, science and opinion;*
> *the former begets knowledge, the latter ignorance.'*

If somebody proposed a new medical treatment, then Hippocrates declared that we should use science to decide whether or not it works, rather than relying on somebody's opinion. Science employs experiments, observations, trials, argument and discussion in order to arrive at an objective consensus on the truth. Even when a conclusion has been decided, science still probes and prods its own proclamations just in case it has made a mistake. In contrast, opinions are subjective and conflicting, and whoever has the most persuasive PR campaign has the best chance of promoting their opinion, regardless of whether they are right or wrong.

Guided by Hippocrates' dictum, this book takes a scientific look at the current plethora of alternative treatments that are rapidly growing in popularity. These treatments are piled high in every pharmacy, written about in every magazine, discussed on millions of web pages and used by billions of people, yet they are regarded with scepticism by many doctors.

Indeed, our definition of an alternative medicine is any therapy that is not accepted by the majority of mainstream doctors, and typically this also means that these alternative therapies have mechanisms that lie outside the current understanding of modern medicine. In the language of science, alternative therapies are said to be biologically implausible.

Nowadays it is common to hear the umbrella term 'complementary and alternative medicine', which correctly implies that sometimes these therapies are used alongside and sometimes instead of conventional medicine. Unfortunately it is a lengthy and clumsy phrase, so in a bid for simplicity we have decided to use the term 'alternative medicine' throughout this book.

Surveys show that in many countries over half the population use alternative medicine in one form or another. Indeed, it is estimated that the annual global spend on all alternative medicines is in the region of £40 billion, making it the fastest-growing area of medical spending. So who is right: the critic who thinks alternative medicine is akin to voodoo, or the mother who entrusts her child's health to alternative medicine? There are three possible answers.

1 Perhaps alternative medicine is entirely useless. Perhaps persuasive marketing has fooled us into believing that alternative medicine works. Alternative therapists might seem like nice people, talking as they do about such appealing concepts as 'nature's wonders' and 'ancient wisdom', but they might be misleading the public – or maybe they are even deluding themselves. They also use impressive buzzwords like holistic, meridians, self-healing and individualized. If we could see past the jargon, then would we realize that alternative medicine is just a scam?

2 Or maybe alternative medicine is overwhelmingly effective. Perhaps the sceptics, including many doctors, have simply failed to recognize the benefits of a more holistic, natural, traditional and spiritual approach to

health. Medicine has never claimed to have all the answers, and over and over again there have been revolutions in our understanding of the human body. So will the next revolution lead to a discovery of the mechanisms underlying alternative medicine? Or could there be darker forces at work? Could it be that the medical establishment wants to maintain its power and authority, and that doctors criticize alternative medicine in order to quash any rivals? Or might these self-same sceptics be puppets of the big pharmaceutical corporations who merely want to hold on to their profits?

3   Or does the truth lie somewhere in the middle?

Whatever the answer, we decided to write this book in order to get to the truth. Although there are already plenty of books that claim to tell you the truth about alternative medicine, we are confident that ours offers an unparalleled level of rigour, authority and independence. We are both trained scientists, so we will examine the various alternative therapies in a scrupulous manner. Moreover, neither of us has ever been employed by a pharmaceutical company, and nor have we ever personally profited from the 'natural health' sector – we can honestly say that our only motive is to get to the truth.

And our partnership brings balance to the book. One of us, Edzard Ernst, is an insider who practised medicine for many years, including some alternative therapies. He is the world's first professor of alternative medicine, and his research group has spent fifteen years trying to work out which treatments work and which do not. The other of us, Simon Singh, is an outsider who has spent almost two decades as a science journalist, working in print, television and radio, always striving to explain complicated ideas in a way that the general public can grasp. Together we think that we can get closer to the truth than anybody else and, equally importantly, we will endeavour to explain it to you in a clear, vivid and comprehensible manner.

Our mission is to reveal the truth about the potions, lotions, pills, needles, pummelling and energizing that lie beyond the realms of conventional medicine, but which are becoming increasingly attractive for many patients. What works, and what doesn't? What are the secrets and what are the lies? Who is trustworthy and who is ripping you off? Do today's doctors know what is best, or do the old wives' tales indeed tap into some ancient, superior wisdom? All these questions and more will be answered in this book, making it the world's most honest and accurate examination of alternative medicine.

In particular, we will answer the fundamental question: 'Is alternative medicine effective for treating disease?' Although a short and simple question, when unpacked it becomes somewhat complicated and has many answers depending on three key issues. First, which alternative therapy are we talking about? Second, which condition are we applying it to? Third, what is meant by effective? In order to address these questions properly, we have divided the book into six chapters.

Chapter 1 provides an introduction to the scientific method. It explains how scientists, by experimenting and observing, can determine whether or not a therapy is effective. Every conclusion we reach in the rest of this book depends on the scientific method and on an unbiased analysis of the best medical research. So, by first explaining how science works, we hope to increase your confidence in our subsequent conclusions.

Chapter 2 shows how the scientific method can be applied to acupuncture, one of the most established, most tested and most widely used alternative therapies. As well as examining the numerous scientific trials that have been conducted on acupuncture, this chapter will also look at its ancient origins in the East, how it migrated to the West and how it is practised today.

Chapters 3, 4 and 5 use a similar approach to examine the three other major alternative therapies, namely homeopathy, chiropractic therapy and herbal medicine. The remaining alternative therapies will be covered in the appendix, which offers a brief analysis of over thirty treatments. In other words, every alternative therapy that you are ever likely to encounter will be scientifically evaluated within the pages of this book.

The sixth and final chapter draws some conclusions based on the evidence in the previous chapters and looks ahead to the future of healthcare. If there is overwhelming evidence that an alternative therapy does not work, then should it be banned or is patient choice the key driving force? On the other hand, if some alternative therapies are genuinely effective, can they be integrated within mainstream medicine or will there always be an antagonism between the establishment and alternative therapists?

The key theme running throughout all six chapters is 'truth'. Chapter 1 discusses how science determines the truth. Chapters 2–5 reveal the truth about various alternative therapies based on the scientific evidence. Chapter 6 looks at why the truth matters, and how this should influence our attitude towards alternative therapies in the context of twenty-first-century medicine.

Truth is certainly a reassuring commodity, but in this book it comes with two warnings. First, we will present the truth in an unapologetically blunt manner. So where we find that a particular therapy does indeed work for a particular illness (e.g. St John's wort does have antidepressive properties, if used appropriately – see Chapter 5), we will say so clearly. In other cases, however, where we discover that a particular therapy is useless, or even harmful, then we shall state this conclusion equally forcefully. You have decided to purchase this book in order to find out the truth, so we think we owe it to you to be direct and honest.

The second warning is that all the truths in this book are based on science, because Hippocrates was absolutely correct when he said that science begets knowledge. Everything we know about the universe, from the components of an atom to the number of galaxies, is thanks to science, and every medical breakthrough, from the development of antiseptics to the eradication of smallpox, has been built upon scientific foundations. Of course, science is not perfect. Scientists will readily admit that they do not know everything, but nevertheless the scientific method is without doubt the best mechanism for getting to the truth.

If you are a reader who is sceptical about the power of science, then we kindly request that you at least read Chapter 1. By the end of that first chapter, you should be sufficiently convinced about the value of scientific method that you will consider accepting the conclusions in the rest of the book.

It could be, however, that you refuse to acknowledge that science is the best way to decide whether or not an alternative therapy works. It might be that you are so close-minded that you will stick to your worldview regardless of what science has to say. You might have an unwavering belief that all alternative medicine is rubbish, or you might adamantly hold the opposite view, that alternative medicine offers a panacea for all our aches, pains and diseases. In either case, this is not the book for you. There is no point in even reading the first chapter if you are not prepared to consider the possibility that the scientific method can act as the arbiter of truth. In fact, if you have already made up your mind about alternative medicine, then it would be sensible for you to return this book to the bookshop and ask for a refund. Why on Earth would you want to hear about the conclusions of thousands of research studies when you already have all the answers?

But our hope is that you will be sufficiently open-minded to want to read further.

1 How Do You Determine the Truth?

'Truth exists – only lies are invented.'

Georges Braque

THIS BOOK IS ABOUT ESTABLISHING THE TRUTH IN RELATION TO alternative medicine. Which therapies work and which ones are useless? Which therapies are safe and which ones are dangerous?

These are questions that doctors have asked themselves for millennia in relation to all forms of medicine, and yet it is only comparatively recently that they have developed an approach that allows them to separate the effective from the ineffective, and the safe from the dangerous. This approach, known as *evidence-based medicine*, has revolutionized medical practice, transforming it from an industry of charlatans and incompetents into a system of healthcare that can deliver such miracles as transplanting kidneys, removing cataracts, combating childhood diseases, eradicating smallpox and saving literally millions of lives each year.

We will employ the principles of evidence-based medicine to test alternative therapies, so it is crucial that we properly explain what it is and how it works. Rather than introducing it in a modern context, we will go back in time to see how it emerged and evolved, which will provide a deeper appreciation of its inherent strengths. In particular, we will look back at how this approach was used to test *bloodletting*, a bizarre and previously common treatment that involved cutting skin and severing blood vessels in order to cure every ailment.

The boom in bloodletting started in Ancient Greece, where it fitted in naturally with the widespread view that diseases were caused by an imbalance of four bodily fluids, otherwise known as the four *humours*: blood, yellow bile, black bile and phlegm. As well as affecting health, imbalances in these humours resulted in particular temperaments. Blood was associated with being optimistic, yellow bile with being irascible, black bile with being depressed and phlegm with being unemotional. We can still hear the echo of humourism in words such as sanguine, choleric, melancholic and phlegmatic.

Unaware of how blood circulates around the body, Greek physicians believed that it could become stagnant and thereby cause ill-health. Hence, they advocated the removal of this stagnant blood, prescribing specific procedures for different illnesses. For example, liver problems were treated by tapping a vein in the right hand, whereas ailments relating to the spleen required tapping a vein in the left hand.

The Greek medical tradition was held in such reverence that bloodletting grew to be a popular method for treating patients throughout Europe in the centuries that followed. Those who could afford it would often receive bloodletting from monks in the early Middle Ages, but then in 1163 Pope Alexander III banned them from practising this gory medical procedure. Thereafter it became common for barbers to take on the responsibility of being the local bleeder. They took their role very seriously, carefully refining their techniques and adopting new technologies. Alongside the simple blade, there was the *phleam*, a spring-loaded blade that cut to a particular depth. In later years this was followed by the *scarificator*, which consisted of a dozen or more spring-loaded blades that simultaneously lacerated the skin.

For those barbers who preferred a less technological and more natural approach, there was the option of using medicinal leeches. The business end of these bloodsucking parasitic worms has three separate jaws, each one of them carrying about 100 delicate teeth. They offered an ideal method for bloodletting from a patient's gums, lips or nose. Moreover, the leech delivers an anaesthetic to reduce pain, an anticoagulant to prevent the blood from clotting, and a vasodilator to expand its victim's blood vessels and increase flow. To enable major bloodsucking sessions, doctors would perform *bdellatomy*, which involved slicing into the leech so that blood entered its sucker end and then leaked out of the cut. This prevented the leech from becoming full and encouraged it to continue sucking.

It is often said that today's red-and-white barbershop pole is emblematic of the barber's earlier role as surgeon, but it is really associated with his position as bleeder. Red represents the blood, white is the tourniquet, the ball at the end symbolizes the brass leech basin and the pole itself represents the stick that was squeezed by the patient to increase blood flow.

Meanwhile, bloodletting was also practised and studied by the most senior medical figures in Europe, such as Ambroise Paré, who was the official royal surgeon to four French kings during the sixteenth century. He wrote extensively on the subject, offering lots of useful hints and tips:

> If the leeches be handled with the bare hand, they are angered, and become so stomachfull as that they will not bite; wherefore you shall hold them in a white and clean linen cloth, and apply them to the skin being first lightly scarified, or besmeared with the blood of some other creature, for thus they will take hold of the flesh, together with the skin more greedily and fully. To cause them to fall off, you shall put some powder of Aloes, salt or ashes upon their heads. If any desire to know how much blood they have drawn, let him sprinkle them with salt made into powder, as soon as they are come off, for thus they will vomit up what blood soever they have sucked.

When Europeans colonized the New World, they took the practice of bloodletting with them. American physicians saw no reason to question the techniques taught by the great European hospitals and universities, so they also considered bloodletting to be a mainstream medical procedure that could be used in a variety of circumstances. However, when it was administered to the nation's most important patient in 1799, its use suddenly became a controversial issue. Was bloodletting really a life-saving medical intervention, or was it draining the life out of patients?

The controversy began on the morning of 13 December 1799, the day that George Washington awoke with the symptoms of a cold. When his personal secretary suggested that he take some medicine, Washington replied, 'You know I never take anything for a cold. I'll let it go just as it came.'

The sixty-seven-year-old former president did not think that a sniffle and a sore throat were anything to worry about, particularly as he had previously suffered and survived far more severe sicknesses. He had contracted smallpox as a teenager, which was followed by a bout of tuberculosis. Next, when he was a young surveyor, he caught malaria while working in the mosquito-infested swamps of Virginia. Then, in 1755, he miraculously survived the Battle of Monongahela, even though two horses were killed beneath him and four musket balls pierced his uniform. He also suffered from pneumonia, was repeatedly afflicted by further bouts of malaria, and developed 'a malignant carbuncle' on his hip that incapacitated him for six weeks.

Perversely, having survived bloody battlefields and dangerous diseases, this apparently minor cold contracted on Friday 13th would prove to be the greatest threat to Washington's life.

His condition deteriorated during Friday night, so much so that he awoke in the early hours gasping for air. When Mr Albin Rawlins, Washington's estate overseer, concocted a mixture of molasses, vinegar and butter, he found that his patient could hardly swallow it. Rawlins, who was also an accomplished bloodletter, decided that further action was required. Anxious to alleviate his master's symptoms, he used a surgical knife known as a lancet to create a small incision in the General's arm and removed one-third of a litre of blood into a porcelain bowl.

By the morning of 14 December there was still no sign of any improvement, so Martha Washington was relieved when three doctors arrived at the house to take care of her husband. Dr James Craik, the General's personal physician, was accompanied by Dr Gustavus Richard Brown and Dr Elisha Cullen Dick. They correctly diagnosed *cynanche trachealis* ('dog strangulation'), which we would today interpret as a swelling and inflammation of the epiglottis. This would have obstructed Washington's throat and led to his difficulty in breathing.

Dr Craik applied some cantharides (a preparation of dried beetles) to his throat. When this did not have any effect, he opted to bleed the General and removed another half a litre of blood. At 11 a.m. he removed a similar amount again. The average human body contains only 5 litres of blood, so a significant fraction was being bled from Washington at each session. Dr Craik did not seem concerned. He performed venesection again in the afternoon, removing a further whole litre of blood.

Over the next few hours, it appeared that the bloodletting was helping. Washington seemed to recover and for a while he was able to sit upright. This was, however, merely a temporary remission. When his condition deteriorated again later that day, the doctors conducted yet another session of bloodletting. This time the blood appeared viscous and flowed slowly. From a modern perspective this reflects dehydration and a general loss of bodily fluid caused by excessive blood loss.

As the evening passed, the doctors could only watch grimly as their numerous bloodlettings and various poultices failed to deliver any signs of recovery. Dr Craick and Dr Dick would later write: 'The powers of life seemed now manifestly yielding to the force of the disorder. Blisters were applied to the extremities, together with a cataplasm of bran and vinegar to the throat.'

George Washington Custis, the dying man's step-grandson, documented the final moments of America's first President:

> As the night advanced it became evident that he was sinking, and he seemed fully aware that 'his hour was nigh'. He inquired the time, and was answered a few minutes to ten. He spoke no more – the hand of death was upon him, and he was conscious that 'his hour was come'. With surprising self-possession he prepared to die. Composing his form at length and folding his arms on his bosom, without a sigh, without a groan, the Father of his Country died. No pang or struggle told when the noble spirit took its noiseless flight; while so tranquil appeared the manly features in the repose of death, that some moments had passed ere those around could believe that the patriarch was no more.

George Washington, a giant man of 6 feet 3½ inches, had been drained of half his blood in less than a day. The doctors responsible for treating Washington claimed that such drastic measures had been necessary as a last-ditch resort to save the patient's life, and most of their colleagues supported the decision. However, there were also voices of dissent from within the medical community. Although bloodletting had been an accepted practice in medicine for centuries, a minority of doctors were now beginning to question its value. Indeed, they argued that bloodletting was a hazard to patients, regardless of where on the body it took place and irrespective of whether it was half a litre or 2 litres that was being taken. According to these doctors, Dr Craik, Dr Brown and Dr Dick had effectively killed the former President by needlessly bleeding him to death.

But who was right – the most eminent doctors in the land who had done their best to save Washington, or the maverick medics who saw bloodletting as a crazy and dangerous legacy of Ancient Greece?

Coincidentally, on the very day that Washington died, 14 December 1799, there was effectively a legal judgement on whether bloodletting was harming or healing patients. The judgement arose as the result of an article by the renowned English journalist William Cobbett, who was living in Philadelphia and who had taken an interest in the activities of a physician by the name of Dr Benjamin Rush, America's most vociferous and famous advocate of bloodletting.

Dr Rush was admired throughout America for his brilliant medical, scientific and political career. He had written eighty-five significant publications, including the first American chemistry textbook; he had been surgeon general of the Continental Army; and, most important of all, he had been a signatory to the Declaration of Independence. Perhaps his achievements were to be expected, bearing in mind that he graduated at the age of just fourteen from the College of New Jersey, which later became Princeton University.

Rush practised at the Pennsylvania Hospital in Philadelphia and taught at its medical school, which was responsible for training three-quarters of American doctors during his tenure. He was so respected that he was known as 'the Pennsylvania Hippocrates' and is still the only physician to have had a statue erected in his honour in Washington DC by the American Medical Association. His prolific career had allowed him to persuade an entire generation of doctors of the benefits of bloodletting, including the three doctors who had attended General Washington. For Rush had served with Dr Craik in the Revolutionary War, he had studied medicine with Dr Brown in Edinburgh, and he had taught Dr Dick in Pennsylvania.

Dr Rush certainly practised what he preached. His best-documented bloodletting sprees took place during the Philadelphia yellow fever epidemics of 1794 and 1797. He sometimes bled 100 patients in a single day, which meant that his clinic had the stench of stale blood and attracted swarms of flies. However, William Cobbett, who had a particular interest in reporting on medical scandals, was convinced that Rush was inadvertently killing many of his patients. Cobbett began examining the local bills of mortality and, sure enough, noticed an increase in death rates after Rush's colleagues followed his recommendations for bloodletting. This prompted him to declare that Rush's methods had 'contributed to the depopulation of the Earth'.

Dr Rush's response to this allegation of malpractice was to sue Cobbett for libel in Philadelphia in 1797. Delays and distractions meant that the case dragged on for over two years, but by the end of 1799 the jury was ready to make a decision. The key issue was whether Cobbett was correct in claiming that Rush was killing his patients through bloodletting, or whether his accusation was unfounded and malicious. While Cobbett could point to the bills of mortality to back up his case, this was hardly a rigorous analysis of the impact of bloodletting. Moreover, everything else was stacked against him.

For example, the trial called just three witnesses, who were all doctors sympathetic to Dr Rush's approach to medicine. Also, the case was argued by seven lawyers, which suggests that powers of persuasion were more influential than evidence. Rush, with his wealth and reputation, had the best lawyers in town arguing his case, so Cobbett was always fighting an uphill battle. On top of all this, the jury was probably also swayed by the fact that Cobbett was not a doctor, whereas Rush was one of the fathers of American medicine, so it would have seemed natural to back Rush's claim.

Not surprisingly, Rush won the case. Cobbett was ordered to pay $5,000 to Rush in compensation, which was the largest award ever paid out in Pennsylvania. So, at exactly the same time that George Washington was dying after a series of bloodletting procedures, a court was deciding that it was a perfectly satisfactory medical treatment.

We cannot, however, rely on an eighteenth-century court to decide whether or not the medical benefits of bloodletting outweigh any damaging side-effects. After all, the judgement was probably heavily biased by all the factors already mentioned. It is also worth remembering that Cobbett was a foreigner, whereas Rush was a national hero, so a judgement against Rush was almost unthinkable.

In order to decide the true value of bloodletting, the medical profession would require a more rigorous procedure, something even less biased than the fairest court imaginable. In fact, while Rush and Cobbett were debating medical matters in a court of law, they were unaware that exactly the right sort of procedure for establishing the truth about medical matters had already been discovered on the other side of the Atlantic and was being used to great effect. Initially it was used to test a radically new treatment for a disease that afflicted only sailors, but it would soon be used to evaluate bloodletting, and in time this approach would be brought to bear on a whole range of medical interventions, including alternative therapies.

Scurvy, limeys and the blood test

In June 1744 a hero of the British navy named Commander George Anson returned home having completed a circumnavigation of the world that had taken almost four years. Along the way, Anson had fought and captured the Spanish galleon *Covadonga*, including its 1,313,843 pieces of eight and

35,682 ounces of virgin silver, the most valuable prize in England's decade of fighting against Spain. When Anson and his men paraded through London, his booty accompanied him in thirty-two wagons filled with bullion. Anson had, however, paid a high price for these spoils of war. His crew had been repeatedly struck by a disease known as *scurvy*, which had killed more than two out of three of his sailors. To put this into context, while only four men had been killed during Anson's naval battles, over 1,000 had succumbed to scurvy.

Scurvy had been a constant curse ever since ships had set sail on voyages lasting for more than just a few weeks. The first recorded case of naval scurvy was in 1497 as Vasco da Gama rounded the Cape of Good Hope, and thereafter the incidences increased as emboldened captains sailed further across the globe. The English surgeon William Clowes, who had served in Queen Elizabeth's fleet, gave a detailed description of the horrendous symptoms that would eventually kill two million sailors:

> Their gums were rotten even to the very roots of their very teeth, and their cheeks hard and swollen, the teeth were loose neere ready to fall out . . . their breath a filthy savour. The legs were feeble and so weak, that they were full of aches and paines, with many blewish and reddish staines or spots, some broad and some small like flea-biting.

All this makes sense from a modern point of view, because we know that scurvy is the result of vitamin C deficiency. The human body uses vitamin C to produce collagen, which glues together the body's muscles, blood vessels and other structures, and so helps to repair cuts and bruises. Hence, a lack of vitamin C results in bleeding and the decay of cartilage, ligaments, tendons, bone, skin, gums and teeth. In short, a scurvy patient disintegrates gradually and dies painfully.

The term 'vitamin' describes an organic nutrient that is vital for survival, but which the body cannot produce itself; so it has to be supplied through food. We typically obtain our vitamin C from fruit, something that was sadly lacking from the average sailor's diet. Instead, sailors ate biscuits, salted meat, dried fish, all of which were devoid of vitamin C and likely to be riddled with weevils. In fact, infestation was generally considered to be a good sign, because the weevils would abandon the meat only when it became dangerously rotten and truly inedible.

The simple solution would have been to alter the sailors' diet, but scientists had yet to discover vitamin C and were unaware of the importance of fresh fruit in preventing scurvy. Instead, physicians proposed a whole series of other remedies. Bloodletting, of course, was always worth a try, and other treatments included the consumption of mercury paste, salt water, vinegar, sulphuric acid, hydrochloric acid or Moselle wine. Another treatment required burying the patient up to his neck in sand, which was not even very practical in the middle of the Pacific. The most twisted remedy was hard labour, because doctors observed that scurvy was generally associated with lazy sailors. Of course, the doctors had confused cause and effect, because it was scurvy that caused sailors to be lazy, rather than laziness that made sailors vulnerable to scurvy.

This array of pointless remedies meant that maritime ambitions during the seventeenth and eighteenth centuries continued to be blighted by deaths from scurvy. Learned men around the world would fabricate arcane theories about the causes of scurvy and debate the merits of various cures, but nobody seemed capable of stopping the rot that was killing hundreds of thousands of sailors. Then, in 1746, there came a major breakthrough when a young Scottish naval surgeon called James Lind boarded HMS *Salisbury*. His sharp brain and meticulous mind allowed him to discard fashion, prejudice, anecdote and hearsay, and instead he tackled the curse of scurvy with extreme logic and rationality. In short, James Lind was destined to succeed where all others had failed because he implemented what seems to have been the world's first *controlled clinical trial.*

Lind's tour of duty took him around the English Channel and Mediterranean, and even though HMS *Salisbury* never strayed far from land, one in ten sailors showed signs of scurvy by the spring of 1747. Lind's first instinct was probably to offer sailors one of the many treatments popular at the time, but this was overtaken by another thought that crossed his mind. What would happen if he treated different sailors in different ways?

By observing who recovered and who deteriorated he would be able to determine which treatments were effective and which were useless. To us this may seem obvious, but it was a truly radical departure from previous medical custom.

On 20 May Lind identified twelve sailors with simlarly serious symptoms of scurvy, inasmuch as they all had 'putrid gums, the spots and lassitude, with weakness of their knees'. He then placed their hammocks in the same portion of the ship and ensured that they all received the same breakfast, lunch and dinner, thereby establishing 'one diet

James Lind

common to all'. In this way, Lind was helping to guarantee a fair test because all the patients were similarly sick, similarly housed and similarly fed.

He then divided the sailors into six pairs and gave each pair a different treatment. The first pair received a quart of cider, the second pair received twenty-five drops of elixir of vitriol (sulphuric acid) three times a day, the third pair received two spoonfuls of vinegar three times a day, the fourth pair received half a pint of sea water a day, the fifth pair received a medicinal paste consisting of garlic, mustard, radish root and gum myrrh, and the sixth pair received two oranges and a lemon each day. Another group of sick sailors who continued with the normal naval diet were also monitored and acted as a control group.

There are two important points to clarify before moving on. First, the inclusion of oranges and lemons was a shot in the dark. Although there had been a few reports of lemons relieving symptoms of scurvy as far back as 1601, late-eighteenth-century doctors would have viewed fruit as a bizarre remedy. Had the term 'alternative medicine' existed in Lind's era, then his colleagues might have labelled oranges and lemons as alternative, as they were natural remedies that were not backed by a plausible theory, and thus they were unlikely to compare well against the more established medicines.

The second important point is that Lind did not include bloodletting in his trial. Although others may have felt that bloodletting was appropriate for treating scurvy, Lind was unconvinced and instead he suspected that the genuine cure would be related to diet. We shall return to the question of testing bloodletting shortly.

The clinicial trial began and Lind waited to see which sailors, if any, would recover. Although the trial was supposed to last fourteen days, the ship's supply of citrus fruits came to an end after just six days, so Lind had to evaluate the results at this early stage. Fortunately, the conclusion was already obvious, for the sailors who were consuming lemons and oranges had made a remarkable and almost complete recovery. All the other patients were still suffering from scurvy, except for the cider drinkers who showed slight signs of improvement. This is probably because cider can also contain small amounts of vitamin C, depending on how it is made.

By controlling variables such as environment and diet, Lind had demonstrated that oranges and lemons were the key to curing scurvy. Whilst the numbers of patients involved in the trial were extremely small, the results he obtained were so striking that he was convinced by the findings. He had no idea, of course, that oranges and lemons contain vitamin C, or that vitamin C is a key ingredient in the production of collagen, but none of this was important – the bottom line was that his treatment led to a cure. Demonstrating that a treatment is effective is the number-one priority in medicine; understanding the exact details of the underlying mechanism can be left as a problem for subsequent research.

Had Lind been researching in the twenty-first century, he would have reported his findings at a major conference and subsequently published them in a medical journal. Other scientists would have read his methodology and repeated his trial, and within a year or two there would have been an international consensus on the ability of oranges and lemons to cure scurvy. Unfortunately, the eighteenth-century medical community was comparatively splintered, so new breakthroughs were often overlooked.

Lind himself did not help matters because he was a diffident man, who failed to publicize and promote his research. Eventually, six years after the trial, he did write up his work in a book dedicated to Commander Anson, who had famously lost over 1,000 men to scurvy just a few years earlier. *Treatise on the Scurvy* was an intimidating tome consisting of 400 pages written in a plodding style, so not surprisingly it won him few supporters.

Worse still, Lind undermined the credibility of his cure with his development of a concentrated version of lemon juice that would be easier to transport, store, preserve and administer. This so-called *rob* was created by heating and evaporating lemon juice, but Lind did not realize that this process destroyed vitamin C, the active ingredient that cured scurvy. Therefore, anybody who followed Lind's recommendation soon became disillusioned, because the lemon rob was almost totally ineffective. So, despite a successful trial, the simple lemon cure was ignored, scurvy continued unabated and many more sailors died. By the time that the Seven Years War with France had ended in 1763, the tallies showed that 1,512 British sailors had been killed in action and 100,000 had been killed by scurvy.

However, in 1780, thirty-three years after the original trial, Lind's work caught the eye of the influential physician Gilbert Blane. Nicknamed 'Chillblain' because of his frosty demeanour, Blane had stumbled upon Lind's treatise on scurvy while he was preparing for his first naval posting with the British fleet in the Caribbean. He was impressed by Lind's declaration that he would 'propose nothing dictated merely from theory; but shall confirm all by experience and facts, the surest and most unerring guides'. Inspired by Lind's approach and interested in his conclusion, Blane decided that he would scrupulously monitor mortality rates throughout the British fleet in the West Indies in order to see what would happen if he introduced lemons to the diet of all sailors.

Although Blane's study was less rigorously controlled than Lind's research, it did involve a much larger number of sailors and its results were arguably even more striking. During his first year in the West Indies there were 12,019 sailors in the

British fleet, of whom only sixty died in combat and a further 1,518 died of disease, with scurvy accounting for the overwhelming majority of these deaths. However, after Blane introduced lemons into the diet, the mortality rate was cut in half. Later, limes were often used instead of lemons, which led to *limeys* as a slang term for British sailors and later for Brits in general.

Not only did Blane become convinced of the importance of fresh fruit, but fifteen years later he was able to implement scurvy prevention throughout the British fleet when he was appointed to the Sick and Hurt Board, which was responsible for determining naval medical procedures. On 5 March 1795 the Board and the Admiralty agreed that sailors' lives would be saved if they were issued a daily ration of just three-quarters of an ounce of lemon juice. Lind had died just one year earlier, but his mission to rid British ships of scurvy had been ably completed by Blane.

The British had been tardy in adopting lemon therapy, as almost half a century had passed since Lind's groundbreaking trial, but many other nations were even tardier. This gave Britain a huge advantage in terms of colonizing distant lands and winning sea battles with its European neighbours. For example, prior to the Battle of Trafalgar in 1805, Napoleon had planned to invade Britain, but he was prevented from doing so by a British naval blockade that trapped his ships in their home ports for several months. Bottling up the French fleet was possible only because the British ships supplied their crews with fruit, which meant that they did not have to interrupt their tour of duty to bring on board new healthy sailors to replace those that would have been dying from scurvy. Indeeed, it is no exaggeration to say that Lind's invention of the clinical trial and Blane's consequent promotion of lemons to treat scurvy saved the nation, because Napoleon's army was much stronger than its British counterpart, so a failed blockade would probably have resulted in a successful French invasion.

The fate of a nation is of major historic importance, yet the application of the clinical trial would have even greater significance in the centuries ahead. Medical researchers would go on to use clinical trials routinely to decide which treatments worked and which were ineffective. In turn, this would allow doctors to save hundreds of millions of lives around the world because they would be able to cure diseases by confidently relying on proven medicines, rather than mistakenly advocating quack remedies.

Bloodletting, because of its central role in medicine, was one of the first treatments to be submitted to testing via the controlled clinical trial. In 1809, just a decade after Washington had undergone bloodletting on his deathbed, a Scottish military surgeon called Alexander Hamilton set out to determine whether or not it was advisable to bleed patients. Ideally, his clinical trial would have examined the impact of bloodletting on a single disease or symptom, such as gonorrhoea or fever, because the results tend to be clearer if a trial is focused on one treatment for one ailment. However, the trial took place while Hamilton was serving in the Peninsular War in Portugal, where battlefield conditions did not afford him the luxury of conducting an ideal trial – instead, he examined the impact of bloodletting on a broad range of conditions. To be fair to Hamilton, this was not such an unreasonable design for his trial, because at the time bloodletting was touted as a panacea – if physicians believed that bloodletting could cure every disease, then it could be argued that the trial should include patients with every disease.

Hamilton began his trial by dividing a sample of 366 soldiers with a variety of medical problems into three groups. The first two groups were treated by himself and a colleague (Mr Anderson) without resorting to bloodletting, whereas the third group was treated by an unnamed doctor who administered the usual treatment of employing a lancet to bleed his patients. The results of the trial were clear:

> It had been so arranged, that this number was admitted, alternately, in such a manner that each of us had one third of the whole. The sick were indiscriminately received, and were attended as nearly as possible with the same care and accommodated with the same comforts . . . Neither Mr Anderson nor I ever once employed the lancet. He lost two, I four cases; whilst out of the other third thirty-five patients died.'

The death rate for patients treated with bloodletting was ten times greater than for those patients who avoided bloodletting. This was a damning indictment on drawing blood and a vivid demonstration that it caused death rather than saved lives. It would have been hard to argue with the trial's conclusion, because it scored highly in terms of two of the main factors that determine the quality of a trial.

First, the trial was carefully controlled, which means that the separate groups of patients were treated similarly except for one particular factor, namely bloodletting. This allowed Hamilton to isolate the impact of bloodletting. Had the bloodletting group been kept in poorer conditions or given a different diet, then the higher death rate could have been attributed to environment or nutrition, but Hamilton had ensured that all the groups received the 'same care' and 'same comforts'. Therefore bloodletting alone could be identified as being responsible for the higher death rate in the third group.

Second, Hamilton had tried to ensure that his trial was fair by guaranteeing that the groups that were being studied were on average as similar as possible. He achieved this by avoiding any systematic assignment of patients, such as deliberately steering elderly soldiers towards the bloodletting group, which would have biased the trial against bloodletting. Instead,

Hamilton assigned patients to each group 'alternately' and 'indiscriminately', which today is known as *randomizing* the allocation of treatments in a trial. If the patients are randomly assigned to groups, then it can be assumed that the groups will be broadly similar in terms of any factor, such as age, income, gender or the severity of the illness, which might affect a patient's outcome. Randomization even allows for unknown factors to be balanced equally across the groups. Fairness through randomization is particularly effective if the initial pool of participants is large. In this case, the number of participants (366 patients) was impressively large. Today medical researchers call this a *randomized controlled trial* (or RCT) or a *randomized clinical trial*, and it is considered the gold standard for putting therapies to the test.

Although Hamilton succeeded in conducting the first randomized clinical trial on the effects of bloodletting, he failed to publish his results. In fact, we know of Hamilton's research only because his documents were rediscovered in 1987 among papers hidden in a trunk lodged with the Royal College of Physicians in Edinburgh. Failure to publish is a serious dereliction of duty for any medical researcher, because publication has two important consequences. First, it encourages others to replicate the research, which might either reveal errors in the original research or confirm the result. Second, publication is the best way to disseminate new research, so that others can apply what has been learned.

Lack of publication meant that Hamilton's bloodletting trial had no impact on the widespread enthusiasm for the practice. Instead, it would take a few more years before other medical pioneers, such as the French doctor Pierre Louis, would conduct their own trials and confirm Hamilton's conclusion. These results, which were properly published and disseminated, repeatedly showed that bloodletting was not a lifesaver, but rather it was a potential killer. In light of these findings, it seems highly likely that bloodletting was largely responsible for the death of George Washington.

Unfortunately, because these anti-bloodletting conclusions were contrary to the prevailing view, many doctors struggled to accept them and even tried their best to undermine them. For example, when Pierre Louis published the results of his trials in 1828, many doctors dismissed his negative conclusion about bloodletting precisely because it was based on the data gathered by analysing large numbers of patients. They slated his so-called 'numerical method' because they were more interested in treating the individual patient lying in front of them than in what might happen to a large sample of patients. Louis responded by arguing that it was impossible to know whether or not a treatment might be safe and effective for the individual patient unless it had been demonstrated to be safe and effective for a large number of patients: 'A therapeutic agent cannot be employed with any discrimination or probability of success in a given case, unless its general efficacy, in analogous cases, has been previously ascertained . . . without the aid of statistics nothing like real medicine is possible.'

And when the Scottish doctor Alexander MacLean advocated the use of medical trials to test treatments while he was working in India in 1818, critics argued that it was wrong to experiment with the health of patients in this way. He responded by pointing out that avoiding trials would mean that medicine would for ever be nothing more than a collection of untested treatments, which might be wholly ineffective or dangerous. He described medicine practised without any evidence as 'a continued series of experiments upon the lives of our fellow creatures.'

Despite the invention of the clinical trial and regardless of the evidence against bloodletting, many European doctors continued to bleed their patients, so much so that France had to import 42 million leeches in 1833. But as each decade passed, rationality began to take hold among doctors, trials became more common, and dangerous and useless therapies such as bloodletting began to decline.

Prior to the clinical trial, a doctor decided his treatment for a particular patient by relying on his own prejudices, or on what he had been taught by his peers, or on his mis-remembered experiences of dealing with a handful of patients with a similar condition. After the advent of the clinical trial, doctors could choose their treatment for a single patient by examining the evidence from several trials, perhaps involving thousands of patients. There was still no guarantee that a treat-ment that had succeeded during a set of trials would cure a particular patient, but any doctor who adopted this approach was giving his patient the best possible chance of recovery.

Lind's invention of the clinical trial had triggered a gradual revolution that gained momentum during the course of the nineteenth century. It transformed medicine from a dangerous lottery in the eighteenth century into a rational discipline in the twentieth century. The clinical trial helped give birth to modern medicine, which has enabled us to live longer, healthier, happier lives.

Evidence-based medicine

Because clinical trials are an important factor in determining the best treatments for patients, they have a central role within a movement known as *evidence-based medicine*. Although the core principles of evidence-based medicine would have been appreciated by James Lind back in the eighteenth century, the movement did not really take hold until the mid-twentieth century, and the term itself did not appear in print until 1992, when it was coined by David Sackett at McMaster University, Ontario. He defined it thus: 'Evidence-based medicine is the con-scientious, explicit, and judicious use of current best evidence in making decisions about the care of individual patients.'

Evidence-based medicine empowers doctors by providing them with the most reliable information, and therefore it benefits patients by increasing the likelihood that they will receive the most appropriate treatment. From a twenty-first-century perspective, it seems obvious that medical decisions should be based on evidence, typically from randomized clinical trials, but the emergence of evidence-based medicine marks a turning point in the history of medicine.

Prior to the development of evidence-based medicine, doctors were spectacularly ineffective. Those patients who recovered from disease were usually successful despite the treatments they had received, not because of them. But once the medical establishment had adopted such simple ideas as the clinical trial, then progress became swift. Today the clinical trial is routine in the development of new treatments and medical experts agree that evidence-based medicine is the key to effective healthcare.

However, people outside the medical establishment sometimes find the concept of evidence-based medicine cold, confusing and intimidating. If you have any sympathy with this point of view, then, once again, it is worth remembering what the world was like before the advent of the clinical trial and evidence-based medicine: doctors were oblivious to the harm they caused by bleeding millions of people, indeed killing many of them, including George Washington. These doctors were not stupid or evil; they merely lacked the knowledge that emerges when medical trials flourish.

Recall Benjamin Rush, for example, the prolific bleeder who sued for libel and won his case on the day that Washington died. He was a brilliant, well-educated man and a compassionate one, who was responsible for recognizing addiction as a medical condition and realizing that alcoholics lose the capacity to control their drinking behaviour. He was also an advocate for women's rights, fought to abolish slavery and campaigned against capital punishment. However, this combination of intelligence and decency was not enough to stop him from killing hundreds of patients by bleeding them to death, and encouraging many of his students to do exactly the same.

Rush was fooled by his respect for ancient ideas coupled with the ad hoc reasons that were invented to justify the use of bloodletting. For example, it would have been easy for Rush to mistake the sedation caused by bloodletting for a genuine improvement, unaware that he was draining the life out of his patients. He was also probably confused by his own memory, selectively remembering those of his patients who survived bleeding and conveniently forgetting those who died. Moreover, Rush would have been tempted to attribute any success to his treatment and to dismiss any failure as the fault of a patient who in any case was destined to die.

Although evidence-based medicine now condemns the sort of bloodletting that Rush indulged in, it is important to point out that evidence-based medicine also means remaining open to new evidence and revising conclusions. For example, thanks to the latest evidence from new trials, bloodletting is once again an acceptable treatment in very specific situations – it has now been demonstrated, for instance, that bloodletting as a last resort can ease the fluid overload caused by heart failure. Similarly, there is now a role for leeches in helping patients recover from some forms of surgery. For example, in 2007 a woman in Yorkshire had leeches placed in her mouth four times a day for a week and a half after having a cancerous tumour removed and her tongue reconstructed. This was because leeches release chemicals that increase blood flow and thus accelerate healing.

Despite being an undoubted force for good, evidence-based medicine is occasionally treated with suspicion. Some people perceive it as being a strategy for allowing the medical establishment to defend its own members and their treatments, while excluding outsiders who offer alternative treatments. In fact, as we have seen already, the opposite is often true, because evidence-based medicine actually allows outsiders to be heard – it endorses any treatment that turns out to be effective, regardless of who is behind it, and regardless of how strange it might appear to be. Lemon juice as a treatment for scurvy was an implausible remedy, but the establishment had to accept it because it was backed up by evidence from trials. Bloodletting, on the other hand, was very much a standard treatment, but the establishment eventually had to reject its own practice because it was undermined by evidence from trials.

There is one episode from the history of medicine that illustrates particularly well how an evidence-based approach forces the medical establishment to accept the conclusions that emerge when medicine is put to the test. Florence Nightingale, the Lady with the Lamp, was a woman with very little reputation, but she still managed to win a bitter argument against the male-dominated medical establishment by arming herself with solid, irrefutable data. Indeed, she can be seen as one of the earliest advocates of evidence-based medicine, and she successfully used it to transform Victorian healthcare.

Florence and her sister were born during an extended and very productive two-year-long Italian honeymoon taken by their parents William and Frances Nightingale. Florence's older sister was born in 1819 and named Parthenope after the city of her birth – Parthenope being the Greek name for Naples. Then Florence was born in the spring of 1820, and she too was named after the city of her birth. It was expected that Florence Nightingale would grow up to live the life of a privileged English Victorian lady, but as a teenager she regularly claimed to hear God's voice guiding her. Hence, it seems that her desire to become a nurse was the result of a 'divine calling'. This distressed her parents, because nurses were generally viewed as being poorly educated, promiscuous and often drunk, but these were exactly the prejudices that Florence was determined to crush.

The prospect of Florence nursing in Britain was already shocking enough, so her parents would have been doubly terrified by her subsequent decision to work in the hospitals of the Crimean War. Florence had read scandalous reports in newspapers such as *The Times*, which highlighted the large number of soldiers who were succumbing to cholera and malaria. She volunteered her services, and by November 1854 Florence was running the Scutari Hospital in Turkey, which was notorious for its filthy wards, dirty beds, blocked sewers and rotten food. It soon became clear to her that the main cause of death was not the wounds suffered by the soldiers, but rather the diseases that ran rife under such squalid conditions. As one official report admitted, 'The wind blew sewer air up the pipes of numerous open privies into the corridors and wards where the sick were lying.'

Nightingale set about transforming the hospital by providing decent food, clean linen, clearing out the drains and opening the windows to let in fresh air. In just one week she removed 215 handcarts of filth, flushed the sewers nineteen times and buried the carcasses of two horses, a cow and four dogs which had been found in the hospital grounds. The officers and doctors who had previously run the institution felt that these changes were an insult to their professionalism and fought her every step of the way, but she pushed ahead regardless. The results seemed to vindicate her methods: in February 1855 the death rate for all admitted soldiers was 43 per cent, but after her reforms it fell dramatically to just 2 per cent in June 1855. When she returned to Britain in the summer of 1856, Nightingale was greeted as a hero, in large part due to *The Times*'s support:

Wherever there is disease in its most dangerous form, and the hand of the spoiler distressingly nigh, there is that incomparable woman sure to be seen; her benignant presence is an influence for good comfort even amid the struggles of expiring nature. She is a 'ministering angel' without any exaggeration in these hospitals, and, as her slender form glides quietly along each corridor, every fellow's face softens with gratitude at the sight of her.

However, there were still many sceptics. The principal medical officer of the army argued that Nightingale's higher survival rates were not necessarily due to her improved hygiene. He pointed out that her apparent success might have been due to treating soldiers with less serious wounds, or maybe they were treated during a period of milder weather, or maybe there was some other factor that had not been taken into account.

Fortunately, as well as being an exceptionally dedicated military nurse, Nightingale was also a brilliant statistician. Her father, William Nightingale, had been broadminded enough to believe that women should be properly educated, so Florence had studied Italian, Latin, Greek, history, and particularly mathematics. In fact, she had received tutoring from some of Britain's finest mathematicians, such as James Sylvester and Arthur Cayley.

So, when she was challenged by the British establishment, she drew upon this mathematical training and used statistical arguments to back her claim that improved hygiene led to higher survival rates. Nightingale had scrupulously compiled detailed records of her patients throughout her time in the Crimea, so she was able to trawl through them and find all sorts of evidence that proved that she was right about the importance of hygiene in healthcare.

For example, to show that the filth at Scutari Hospital had

been killing soldiers, she used her records to compare a group of soldiers treated at Scutari in the early unhygienic days with a control group of injured soldiers who at the same time were being kept at their own army camp. If the camp-based control group fared better than the Scutari group, then this would indicate that the conditions that Nightingale encountered when she arrived at Scutari were indeed doing more harm than good. Sure enough, the camp-based soldiers had a mortality rate of 27 deaths per 1,000 compared with 427 per 1,000 at Scutari. This was only one set of statistics, but when put alongside other comparisons it helped Nightingale to win her argument about the importance of hygiene.

Nightingale was convinced that all other major medical decisions ought to be based on similar sorts of evidence, so she fought for the establishment of a Royal Commission on the Health of the Army, to which she herself submitted several hundred pages of detailed statistics. At a time when it was considered radical merely to include data tables, she also drew multicoloured diagrams that would not look out of place in a modern boardroom presentation. She even invented an elaborate version of the pie chart, known as the polar area chart, which helped to illustrate her data. She realized that illustrating her statistics would be enormously helpful in selling her argument to politicians, who were usually not well versed in mathematics.

In due course, Nightingale's statistical studies spearheaded a revolution in army hospitals, because the Royal Commission's report led to the establishment of an Army Medical School and a system of collecting medical records. In turn, this resulted in a careful monitoring of which conditions and treatments did and did not benefit patients.

Today, Florence Nightingale is best known as the founder of modern nursing, having established a curriculum and training college for nurses. However, it can be argued that her lifelong campaigning for health reforms based on statistical evidence had an even more significant impact on healthcare. She was elected the first female member of the Royal Statistical Society in 1858, and went on to become an honorary member of the American Statistical Association.

Nightingale's passion for statistics enabled her to persuade the government of the importance of a whole series of health reforms. For example, many people had argued that training nurses was a waste of time, because patients cared for by trained nurses actually had a higher mortality rate than those treated by untrained staff. Nightingale, however, pointed out that this was only because more serious cases were being sent to those wards with trained nurses. If the intention is to compare the results from two groups, then it is essential (as discussed earlier) to assign patients randomly to the two groups. Sure enough, when Nightingale set up trials in which patients were randomly assigned to trained and untrained nurses, it became clear that the cohort of patients treated by trained nurses fared much better than their counterparts in wards with untrained nurses. Furthermore, Nightingale used statistics to show that home births were safer than hospital births, presumably because British homes were cleaner than Victorian hospitals. Her interests also ranged overseas, because she also used mathematics to study the influence of sanitation on healthcare in rural India.

And throughout her long career, Nightingale's commitment to working with soldiers never waned. In one of her later studies, she observed that soldiers based in Britain in peacetime had an annual mortality rate of 20 per 1,000, nearly twice that of civilians, which she suspected was due to poor conditions in their barracks. She calculated the death toll across the whole British army due to poor accommodation and then made a comment that highlighted how this was such a needless waste of young lives: 'You might as well take 1,100

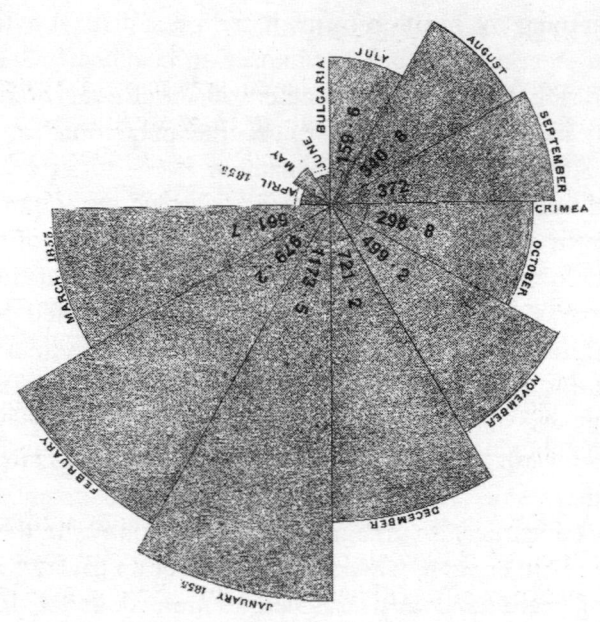

Florence Nightingale's polar chart showing 'Mortality of the Army in the East',
April 1854–March 1855.

men every year out upon Salisbury Plain and shoot them.'

The lesson to be learned from Florence Nightingale's medical triumphs is that scientific testing is not just the best way to establish truth in medicine, but it is also the best mechanism for having that truth recognized. The results from scientific tests are so powerful that they even enable a relative unknown such as Nightingale – a young woman, not part of the establishment, without a great reputation – to prove that she is right and that those in power are wrong. Without medical testing, lone visionaries such as Nightingale would be ignored, while doctors would continue to operate according to a corrupt body of medical knowledge based merely on tradition, dogma, fashion, politics, marketing and anecdote.

A stroke of genius

Before applying an evidence-based approach to evaluating alternative medicine, it is worth re-emphasizing that it provides extraordinarily powerful and persuasive conclusions. Indeed, it is not just the medical establishment that has to tug its forelock in the face of evidence-based medicine, because governments can also be forced to change their policies and corporations may have to alter their products according to what the scientific evidence shows. One final story illustrates exactly how scientific evidence can make the world sit up, listen and act regarding health issues – it concerns the research that dramatically revealed the previously unknown dangers of smoking.

This research was conducted by Sir Austin Bradford Hill and Sir Richard Doll, who had curiously mirrored each other in their backgrounds. Hill had wanted to follow in his father's footsteps and become a doctor, but a bout of tuberculosis made this impossible, so instead he pursued a more mathematical career. Doll's ambition was to study mathematics at Cambridge, but he got drunk on three pints of Trinity Audit Ale (8 per cent alcohol) the night before his entrance exam and underperformed, so instead he pursued a career in medicine. The result was a pair of men with strong interests in both healthcare and statistics.

Hill's career had involved research into a wide variety of health issues. In the 1940s, for instance, he demonstrated a link between arsenic and cancer in chemical workers by examining death certificates, and he went on to prove that rubella during pregnancy could lead to deformities in babies. He also conducted important research into the effectiveness of antibiotics against tuberculosis, the disease that had ended his hopes of becoming a doctor. Then, in 1948, Hill's interest turned towards lung cancer, because there had been a sixfold increase in cases of the disease in just two decades. Experts were divided as to what was behind this health crisis, with some of them dismissing it as a consequence of better diagnosis, while others suggested that lung cancer was being triggered by industrial pollution, car fumes or perhaps smoking.

With no consensus in sight, Hill teamed up with Doll and decided to investigate one of the proposed causes of lung cancer, namely smoking. However, they faced an obvious problem – they could not conduct a randomized clinical trial in this particular context. For instance, it would have been unethical, impractical and pointless to take 100 teenagers, persuade half of them to smoke for a week, and then look for signs of lung cancer.

Instead, Hill and Doll decided that it would be necessary to devise a *prospective cohort study* or an *observational study*, which means that a group of healthy individuals is initially identified and then their subsequent health is monitored while they carry on their day-to-day lives. This is a much less interventionist approach than a randomized clinical trial, which is why a prospective cohort study is preferable for exploring long-term health issues.

To spot any link between smoking and lung cancer in their prospective cohort study, Hill and Doll worked out that they would need to recruit volunteers who fulfilled three important criteria. First, the participants had to be established smokers or vehement non-smokers, because this increased the likelihood that the pattern of behaviour of any individual would continue throughout the study, which would last several years. Second, the participants had to be reliable and dedicated, inasmuch as they would have to commit to the project and submit regular updates on their health and smoking habits during the course of the prospective cohort study. Third, in order to control for other factors, it would help if all the participants were similar in terms of their backgrounds, income and working conditions.

Also, the number of participants had to be large, possibly several thousand, because this would lead to more accurate conclusions.

Finding a group of participants that met these demanding requirements was not a trivial task, but Hill eventually thought of a solution while playing golf. This prompted his friend Dr Wynne Griffith to comment, 'I don't know what kind of golfer he [is], but that was a stroke of genius.' Hill's brilliant idea was to use doctors as his guinea pigs. Doctors fitted the bill perfectly: there were lots of them, many were heavy smokers, they were perfectly able to monitor their health and report back, and they were a relatively homogenous subset of the population.

When the smoking study commenced in 1951, the plan was to monitor more than 30,000 British doctors over the course of five decades, but a clear pattern was already emerging by 1954. There had been thirty-seven deaths from lung cancer and every single one of them was a smoker. As the data accumulated, the study implied that smoking increased the risk of lung cancer by a factor of twenty, and moreover it was linked to a range of other health problems, including heart attacks.

The British Doctors Study, as it was known, was giving rise to such shocking results that some medical researchers were initially reluctant to accept the findings. Similarly, the cigarette industry questioned the research methodology, arguing that there must be a flaw in the way that the information was being gathered or analysed. Fortunately, British doctors were less sceptical about Hill and Doll's emerging conclusions, because they themselves had been so involved in the study. Hence, they were not slow in advising the public against smoking.

Because a link between cigarettes and lung cancer would affect smokers around the world, it was important that the work of Hill and Doll was replicated and checked. The results of another study, this time involving 190,000 Americans, were also announced in 1954, and the conclusion painted a similarly stark picture. Meanwhile, research with mice showed that half of them developed cancerous lesions when their skin was coated in the tarry liquid extracted from tobacco smoke, showing that cigarettes definitely contained carcinogens. The picture was completed with more data from Hill and Doll's ongoing fifty-year study – it reinforced in explicit detail the deadly effects of tobacco. For example, the analysis of British doctors showed that those born in the 1920s who smoked were three times more likely to die in their middle age than their non-smoking colleagues. More specifically, 43 per cent of smokers compared to only 15 per cent of non-smokers died between the ages of 35 and 69 years.

Doll was as shocked as anyone by the damning evidence against smoking: 'I myself did not expect to find smoking was a major problem. If I'd had to bet money at that time, I would have put it on something to do with the roads and motorcars'. Doll and Hill did not start their research in order to achieve a specific result, but instead they were merely curious and concerned about getting to the truth. More generally, well-designed scientific studies and trials are not engineered to achieve an expected outcome, but rather they should be transparent and fair, and those conducting the research should be open to whatever results emerge.

The British Doctors Survey and similar studies were attacked by the tobacco industry, but Doll, Hill and their colleagues fought back and demonstrated that rigorous scientific research can establish the truth with such a level of authority that even the most powerful organizations cannot deny the facts for long. The link between smoking and lung cancer was proved beyond all reasonable doubt because of evidence emerging from several independent sources, each one confirming the results of the other. It is worth reiterating that progress in medicine requires independent replication – i.e. similar studies by more than one research group showing similar findings. Any conclusion that emerges from such a body of evidence is likely to be robust.

Hill and Doll's research ultimately led to a raft of measures designed to persuade us not to smoke, which in turn has resulted in a 50 per cent decrease in smoking in many parts of the developed world. Unfortunately, smoking still remains the single biggest cause of preventable deaths worldwide, because significant new markets are opening up in the developing world. Also, for many smokers the addiction is so great that they ignore or deny the scientific evidence. When Hill and Doll first published their research in the *British Medical Journal*, an accompanying editorial recounted a very telling anecdote: 'It is said that the reader of an American magazine was so disturbed by an article on the subject of smoking and cancer that he decided to give up reading.'

While we were writing this book, the *British Medical Journal* reminded the world of the contribution made by Hill and Doll – it named the research that established the risks of smoking among a list of the fifteen greatest medical break-throughs since the journal was launched 166 years ago. Readers had been asked to vote for their favourite break-through in what seemed like the medical equivalent of *Pop Idol*. Although this high-profile popularity contest might have seemed vulgar to some academics, it made two important points, particularly in the context of this chapter.

First, every breakthrough on the list illustrated the power of science to improve and save lives. For example, the list included oral rehydration, which helps recovery from diarrhoea and which has saved 50 million children's lives in the last twenty-five years. The list also included antibiotics, germ theory and immunology, which together have helped to cure a whole range of diseases, thereby saving hundreds of millions of lives. Vaccines, of course, were on the list, because they have prevented many diseases from even occurring, thereby saving hundreds of millions more lives. And awareness of the risks of smoking has probably saved a similar number of lives.

The second point is that the concept of evidence-based medicine was also recognized among the top fifteen breakthroughs, because it too is a truly great medical achievement. As mentioned earlier, evidence-based medicine is simply about deciding best medical practice based on the best available evidence. It lacks the glamour and glitz of some of the other shortlisted breakthroughs, but it is arguably the greatest one because it underpins so many of the others. For example, the knowledge that vaccines and antibiotics are safe and protect against disease is only possible thanks to evidence gathered through clinical trials and other scientific investigations. Without evidence-based medicine, we risk falling into the trap of considering useless treatments as helpful, or helpful treatments as useless. Without evidence-based medicine we are likely to ignore the best treatments and instead rely on treatments that are mediocre, or poor, or useless, or even dangerous, thereby increasing the suffering of patients.

Even before the principles of evidence-based medicine were formalized, Lind, Hamilton, Louis, Nightingale, Hill and Doll, and hundreds of other medical researchers used the same approach to decide what works (lemons for scurvy), what does not work (bloodletting), what prevents disease (hygiene) and what triggers disease (smoking). The entire framework of modern medicine has emerged thanks to these medical researchers who used scientific methods such as clinical trials to gather evidence in order to get to the truth. Now we can find out what happens when this approach is applied to alternative medicine.

Alternative medicine claims to be able to treat the same ill-nesses and diseases that conventional medicine tries to tackle, and we can test these claims by evaluating the evidence. Any alternative treatment that turns out to be effective for a par-ticular condition can then be compared with conventional medicines to decide if the alternative should be used partially or wholly to replace the conventional.

We are confident that we will be able to offer reliable conclusions about the value of the various alternative therapies, because many researchers have already been con-ducting trials and gathering evidence. In fact, there have been thousands of clinical trials to determine the efficacy of alternative therapies.

Some of them have been conducted with great rigour on large populations of patients and then independently replicated, so the overall conclusions can be relied upon. The remaining chapters of this book are devoted to analysing the results of these trials across a whole range of alternative therapies. Our goal is to examine the evidence and then tell you which therapies work and which ones fail, which ones are safe and which ones are dangerous.

At this early stage of the book, many alternative therapists might feel optimistic that their particular therapy will emerge triumphant when we analyse the data concerning its efficacy. After all, acupuncturists, homeopaths, chiropractors, herbalists and other alternative therapists can probably identify with the mavericks that have populated this chapter.

Florence Nightingale would have been perceived as a maverick during her early career, because she was prioritizing hygiene when everybody else involved in healthcare was focused on other things, such as surgery and pills. But she proved that she was right and that the establishment was wrong.

James Lind was also a maverick who turned out to be right,

because he showed that lemons were effective for scurvy when the medical establishment was promoting all sorts of other remedies. Alexander Hamilton was another maverick who knew more than the establishment, because he argued against bloodletting in an era when bleeding was a standard procedure. And Sir Austin Bradford Hill and Sir Richard Doll were mavericks, because they showed that smoking was a surprisingly deadly indulgence, and moreover they produced data that stood up against the powerful interests of the cigarette industry.

Such heroic mavericks pepper the history of medicine and they also act as powerful role models for modern mavericks, including alternative therapists. Acupuncturists, homeopaths and other practitioners rail against the establishment with theories and therapies that run counter to our current understanding of medicine, and they loudly proclaim that the establishment does not understand them. These therapists predict that, one day, the establishment will acknowledge their apparently strange ideas. They believe that they will earn their own rightful place in the history books, alongside Nightingale, Lind, Hamilton, Hill and Doll. Unfortunately, these alternative therapists ought to realize that only a minority of mavericks ever turn out to be on the right track. Most mavericks are simply deluded and wrong.

Alternative therapists might be excited by a line from George Bernard Shaw's play *Annajanska, the Bolshevik Empress*, in which the Grand Duchess points out: 'All great truths begin as blasphemies.' However, they might be less encouraged by the caveat that should accompany this line: 'Not all blasphemies become great truths.'

Perhaps one of the best reasons to categorize a medical treatment as alternative is if the establishment views it as blasphemous. In this context, the aim of our book is to evaluate the scientific evidence that relates to each alternative treatment to see if it is a blasphemy on the path to revolution-izing medicine or if it is a blasphemy that is destined to remain in the cul-de-sac of crazy ideas.

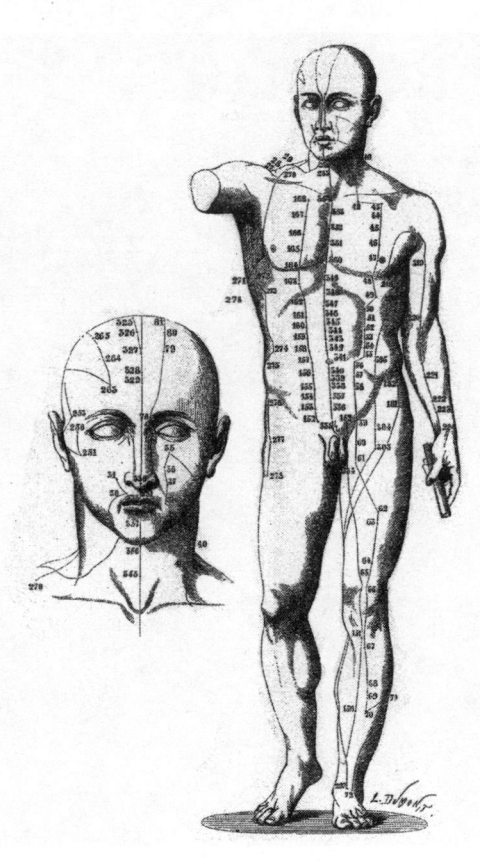

2 The Truth About Acupuncture

*'There must be something to acupuncture
– you never see any sick porcupines.'*

Bob Goddard

Acupuncture

An ancient system of medicine based on the notion that health and wellbeing relate to the flow of a life force (Ch'i) through pathways (meridians) in the human body. Acupuncturists place fine needles into the skin at critical points along the meridians to remove blockages and encourage a balanced flow of the life force. They claim to be able to treat a wide range of diseases and symptoms.

MOST PEOPLE ASSUME THAT ACUPUNCTURE, THE PROCESS OF puncturing the skin with needles to improve health, is a system of medicine that has its origins in China. In fact, the oldest evidence for this practice has been discovered in the heart of Europe. In 1991 two German tourists, Helmut and Erika Simon, were hiking across an alpine glacier in the Ötz valley near the border between Italy and Austria when they encountered a frozen corpse. At first they assumed it was the body of a modern hiker, many of whom have lost their lives due to treacherous weather conditions. In fact, they had stumbled upon the remains of a 5,000-year-old man.

Ötzi the Iceman, named after the valley in which he was found, became world famous because his body had been remarkably well preserved by the intense cold, making him the oldest European mummified human by far. Scientists began examining Ötzi, and soon a startling series of discoveries emerged. The contents of his stomach, for instance, revealed that he had chamois and red-deer meat for his final meals. And, by examining pollen grains mixed in with the meat, it was possible to show that he had died in the spring. He carried with him an axe made of 99.7 per cent pure copper, and his hair showed high levels of copper contamination, implying that he may have smelted copper for a living.

One of the more unexpected avenues of research was initiated by Dr Frank Bahr from the German Academy for Acupuncture and Auriculomedicine. For him, the most interesting aspect of Ötzi was a series of tattoos that covered parts of his body. These tattoos consisted of lines and dots, as opposed to being pictorial, and seemed to form fifteen distinct groups. Moreover, Bahr noticed that the markings were in familiar positions: 'I was amazed – 80 per cent of the points correspond to those used in acupuncture today.'

When he showed the images to other acupuncture experts, they agreed that the majority of tattoos seemed to lie within 6mm of known acupuncture points, and that the remainder all lay close to other areas of special significance to acupuncture. Allowing for the distortion of Ötzi's skin in the past 5,000 years, it was even possible that every single tattoo corresponded with an acupuncture point. Bahr came to the conclusion that the markings were made by an ancient healer in order to allow Ötzi to treat himself by using the tattoos as a guide for applying needles to the correct sites.

Whilst critics have suggested that the overlap between the tattoos and acupuncture points is nothing more than a meaningless coincidence, Bahr remains confident that Ötzi was indeed a prehistoric acupuncture patient. He points out that the pattern of tattoos indicates a particular acupuncture therapy – the majority of tattoo sites are exactly those that would be used by a modern acupuncturist to treat back pain, and the remainder can be linked to abdominal disorders. In a paper published in 1999 in the highly respected journal *Lancet*, Bahr and his colleagues wrote: 'From an acupuncturist's viewpoint, the combination of points selected represents a meaningful therapeutic regimen.' Not only do we have an apparent treatment regime, but we also have a diagnosis that fits the speculation, because radiological studies have shown that Ötzi suffered from arthritis in the lumbar region of his spine, and we also know that there were numerous whipworm eggs in his colon that would have caused him serious abdominal problems.

Despite claims that Ötzi is the world's earliest known acupuncture patient, the Chinese insist that the practice originated in the Far East. According to legend, the effects of acupuncture were serendipitously discovered when a soldier fighting in the Mongolian War in 2,600 BC was struck by an arrow. Fortunately it was not a lethal shot, and even more fortunately it supposedly cured him of a longstanding illness. More concrete evidence for the origins of acupuncture has been found in prehistoric burial tombs, where archaeologists have discovered fine stone tools apparently intended for needling. One line of speculation is that such tools were fashioned because of the ancient Chinese belief that all disease was caused by demons within the human body. It may have been thought that the insertion of needles into the body could kill or release such demons.

The first detailed description of acupuncture appears in the *Huangdi Neijing* (known as the *The Yellow Emperor's Classic of Internal Medicine*), a collection of writings dating from the second century BC. It presents the complex philosophy and practice of acupuncture in terms that would be largely familiar to any modern practitioner. Most importantly of all, *Huangdi Neijing* describes how *Ch'i*, a vital energy or life force, flows though our body via channels known as *meridians*. Illnesses are due to imbalances or blockages in the flow of Ch'i, and the goal of acupuncture is to tap into

Acupuncture chart from the Ming dynasty, showing the meridians, or channels, through which our Ch'i is said to flow.

the meridians at key points to rebalance or unblock the Ch'i.

Although Ch'i is a core principle in acupuncture, different schools have evolved over the centuries and developed their own interpretations of how Ch'i flows through the body. For instance, some acupuncturists work on the basis of fourteen main meridians carrying Ch'i, while the majority support the notion that the body contains only twelve main meridians. Similarly, different schools of acupuncture have included additional concepts, such as yin and yang, and interpreted them in different ways. While some schools divided yin and yang into three subcategories, others divided them into four. Because there are so many schools of acupuncture, it is impractical to give a detailed description of each of them, but these are the core principles:

- Each meridian is associated with and connects to one of the major organs.
- Each meridian has an internal and an external pathway. Although the internal pathways are buried deep within the body, the external ones are relatively near the surface and are accessible to needling.
- There are hundreds of possible acupuncture points along the meridians.
- Depending on the school and the condition being treated, the acupuncturist will insert needles at particular points on particular meridians.
- The penetration depth varies from 1 centimetre to over 10 centimetres, and often the therapy involves rotating the needles in situ.
- Needles can be left in place for a few seconds or a few hours.

Before deciding on the acupuncture points, as well as the duration, depth and mode of needling, the acupuncturist must first diagnose the patient. This relies on five techniques, namely *inspection*, *auscultation*, *olfaction*, *palpation* and *inquiring*. Inspection means examining the body and face, including the colour and coating of the tongue. Auscultation and olfaction entail listening to and smelling the body, checking for symptoms such as wheezing and unusual odours. Palpation involves checking the patient's pulse: importantly, acupuncturists claim to be able to discern far more information from this process than any conventional doctor. Inquiring, as the name suggests, means simply interviewing the patient.

Claims by the Chinese that acupuncture could successfully diagnose and miraculously cure a whole range of diseases inevitably aroused interest from the rest of the world. The first detailed description by a European physician was by Wilhelm ten Rhyne of the Dutch East India Company in 1683, who invented the word acupuncture in his Latin treatise *De Acupunctura*. A few years later, a German traveller and doctor named Engelbert Kaempfer brought back reports of acupuncture from Japan, where it was not restricted to specialist practitioners: 'Even the common people will venture to apply the needle merely upon their own experience . . . taking care only not to prick any nerves, tendons, or considerable blood vessels.'

In time, some European doctors began to practise acupuncture, but they tended to reinterpret the underlying principles to fit in with the latest scientific discoveries. For example, in the early nineteenth century Louis Berlioz, father of the famous composer, found acupuncture to be beneficial for relieving muscular pain and nervous conditions. He speculated that the healing mechanism might be linked to the findings of Luigi Galvani, who had discovered that electrical impulses could cause a dissected frog's leg to twitch. Berlioz suggested that acupuncture needles might be interrupting or enabling the flow of electricity within the body, thereby replacing the abstract notions of Ch'i and meridians with the more tangible concepts of electricity and nerves. This led to Berlioz's proposal that the effects of acupuncture might be enhanced by connecting the needles to a battery.

At the same time, acupuncture was also growing in popularity in America, which prompted some physicians to conduct tests into its efficacy. For example, in 1826 there was an attempt in Philadelphia to resuscitate drowned kittens by inserting needles into their hearts, an experiment based on the claims of European acupuncturists. Unfortunately the American doctors had no success and 'gave up in disgust'.

Meanwhile, European acupuncturists continued to publish articles reporting positive results, such as one that appeared in the *Lancet* in 1836 describing how acupuncture had been used to cure a swelling of the scrotum. At the same time, the therapy

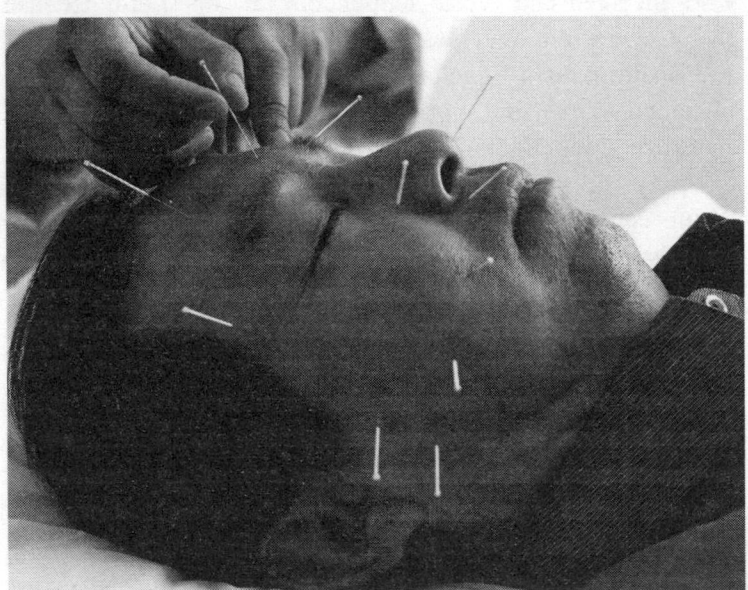

Figure 1
A patient receiving acupuncture at several points on the face. Three yang meridians are said to begin in the facial region.

became particularly popular in high society thanks to its promotion by figures such as George O'Brien, 3rd Earl of Egremont, who was cured of sciatica. He was so impressed that he renamed his favourite racehorse Acupuncture as an act of gratitude towards the wondrous therapy.

Then, from around 1840, just when it seemed that acupuncture was going to establish itself within mainstream Western medicine, the wealthy elite adopted new medical fads and the number of acupuncturists dwindled. European rejection of the practice was mainly linked to disputes such as the First and Second Opium Wars between Britain and China, which led to a contempt for China and its traditions – acupuncture was no longer perceived as a potent therapy from the mysterious East, but instead it was considered a sinister ritual from the evil Orient. Meanwhile, acupuncture was also in decline back in China. The Daoguang Emperor (1782–1850) felt it was a barrier to medical progress and removed it from the curriculum of the Imperial Medical Institute.

By the start of the twentieth century, acupuncture was extinct in the West and dormant in the East. It might have fallen out of favour permanently, but it suddenly experienced a revival in 1949 as a direct result of the communist revolution and the establishment of the People's Republic of China. Chairman Mao Tse-tung engineered a resurgence in traditional Chinese medicine, which included not just acupuncture, but also Chinese herbal medicine and other therapies. His motivation was partly ideological, inasmuch as he wanted to reinforce a sense of national pride in Chinese medicine. However, he was also driven by necessity. He had promised to deliver affordable healthcare in both urban and rural regions, which was only achievable via the network of traditional healers, the so-called 'barefoot doctors'. Mao did not care whether traditional Chinese medicine worked, as long as he could keep the masses contented. In fact, his personal physician, Zhisui Li, wrote a memoir entitled *The*

Private Life of Chairman Mao, in which he quoted Mao as saying, 'Even though I believe we should promote Chinese medicine, I personally do not believe in it. I don't take Chinese medicine.'

Because of China's isolation, its renewed interest in acupuncture went largely unnoticed in the West – a situation which changed only when plans were being made for President Nixon's historic trip to China in 1972. This was the first time that an American President had visited the People's Republic of China, so it was preceded by a preparatory visit by Henry Kissinger in July 1971. Even Kissinger's visit was a major event, so he was accompanied by a cohort of journalists, which included a reporter called James Reston. Unfortunately for Reston, soon after arriving in China he was struck by a stabbing pain in his groin. He later recalled how his condition deteriorated during the day: 'By evening I had a temperature of 103 degrees and in my delirium I could see Mr. Kissinger floating across my bedroom ceiling grinning at me out of the corner of a hooded rickshaw.'

It soon became obvious that he had appendicitis, so Reston was urgently admitted to the Anti-Imperialist Hospital for a standard surgical procedure. The operation went smoothly, but two nights later Reston began to suffer severe abdominal pains which were treated with acupuncture. He was cared for by Dr Li Chang-yuan, who had not been to medical college, but who instead had served an apprenticeship with a veteran acupuncturist. He told Reston that he had learned much of his craft by practising on himself: 'It is better to wound yourself a thousand times than to do a single harm to another person.'

James Reston found the treatment to be both shocking and effective in equal measure, and he wrote up his experience in an article published in the *New York Times* on 26 July 1971. Under the headline 'NOW ABOUT MY OPERATION IN PEKING', Reston described how the acupuncturist had inserted needles into his right elbow and just below both knees. Americans must have been amazed to read how the needles were then 'manipulated in order to stimulate the intestine and relieve the pressure and distension of the stomach'. Reston praised the way that this traditional technique had eased his pain, which resulted in the article generating enormous interest among medical experts. Indeed, it was not long before White House physicians and other American doctors were visiting China to see the power of acupuncture with their own eyes.

During the early 1970s, these observers witnessed truly staggering examples of Chinese acupuncture. Perhaps the most impressive demonstration was the use of acupuncture during major surgery. A certain Dr Isadore Rosenfeld, for instance, visited the hospital at the University of Shanghai and reported on the case of a twenty-eight-year-old female patient who underwent open-heart surgery to repair her mitral valve. Astonishingly, the surgeons used acupuncture to her left earlobe in place of the usual anaesthetics. The surgeon cut through the breastbone with an electric buzzsaw and opened her chest to reveal her heart. Dr Rosenfeld described how she remained awake and alert throughout: 'She never flinched. There was no mask on her face, no intravenous needle in her arm . . . I took a color photograph of that memorable scene: the open chest, the smiling patient, and the surgeon's hands holding her heart. I show it to anyone who scoffs at acupuncture.'

Such extraordinary cases, documented by reputable doctors, had an immediate effect back in America. Physicians were clamouring to attend the three-day crash courses in acupuncture that were running in both America and China, and increasing numbers of acupuncture needles were being imported into America. At the same time, American legislators were deciding what to make of this newfound medical marvel, because there had been no formal assessment of whether or not acupuncture really worked. Similarly there had been no investigation into the safety of acupuncture implements, which was why the Food and Drug Administration (FDA) attempted to prevent shipments of needles from entering the United States. Eventually the FDA softened its position and accepted the importation of acupuncture needles under the label of experimental devices. The Governor of California, Ronald Reagan, took a similar line, and in August 1972 he signed into law a bill that permitted acupuncture, but only in approved medical schools and only so that scientists might test its safety and efficacy.

In hindsight, we can see that those who argued for caution were probably correct. It now seems highly likely that many of the Chinese demonstrations involving surgery had been faked, inasmuch as the acupuncture was being supplemented by local anaesthetics, sedatives or other means of pain control. Indeed, it is a deception that has occurred as recently as 2006, when the BBC TV series *Alternative Medicine* generated national interest after showing an operation that was almost identical to the one observed by Dr Rosenfeld three decades earlier. Again, acupuncture was being used on a female patient in her twenties, also undergoing open-heart surgery, and also in Shanghai.

The BBC's presenter explained that: 'She's still conscious, because instead of a general anaesthetic this twenty-first-century surgical team are using a two-thousand-year-old method of controlling pain – acupuncture.' British journalists and the general public were amazed by the extraordinary images, but a report by the Royal College of Anaesthetists cast the operation in a different light:

It is obvious, from her appearance, that the patient has already received sedative drugs and I am informed that these comprised midazolam, droperidol and fentanyl. The doses used were small, but these types of drugs 'amplify' the effect of each other so that the effect becomes greater. Fentanyl is not actually a sedative drug in the strict sense, but it is a pain-killing drug that is considerably more powerful than morphine. The third component of the anaesthetic is seen on the tape as well, and that is the infiltration of quite large volumes of local anaesthetic into the tissues on the front of the chest where the surgical incision is made.

In short, the patient had received sufficiently large doses of conventional drugs to mean that the acupuncture needles were a red herring, probably playing nothing more than a cosmetic or psychological role.

The American physicians who visited China in the early 1970s were not accustomed to deception or political manipulation, so it took a couple of years before their naïve zeal for acupuncture turned to doubt. Eventually, by the mid-1970s, it had become clear to many of them that the use of acupuncture as a surgical anaesthetic in China had to be treated with scepticism. Films of impressive medical procedures made by the Shanghai Film Studio, which had once been shown in American medical schools, were reinterpreted as propaganda. Meanwhile, the Chinese authorities continued to make outrageous claims for acupuncture, publishing brochures that contained assertions such as: 'Deep needling of the *yamen* point enables deaf-mutes to hear and speak . . . And when the devil was cast out, the dumb spake: and the multitudes marvelled.'

Acupuncture's reputation in the West had risen and fallen in less than a decade. It had been praised unreservedly following President Nixon's visit to China, only later to be treated with suspicion by the medical establishment. This did not mean, however, that Western physicians were necessarily close-minded to the whole notion of acupuncture. The more extraordinary claims might have been unjustified, but perhaps many of the other supposed benefits were genuine. The only way to find out would be for acupuncture to pass through the same protocols that would be required of any new treatment. The situation was best summarized by the American Society of Anesthesiologists, who issued a statement in 1973 that highlighted the need for caution, while also offering a way forward:

> The safety of American medicine has been built on the scientific evaluation of each technique before it becomes a widely accepted concept in medical practice. The premature use of acupuncture in the United States at this time departs from this traditional approach. A potentially valuable technique which has been developed over thousands of years in China is being hastily applied with little thought to safeguards or hazards. Among the potential hazards is the application to the patient who has not been properly evaluated psychologically. If acupuncture is applied indiscriminately, severe mental trauma could result in certain patients. Another hazard is the possible misuse by quacks in attempting to treat a variety of illnesses, including cancer and arthritis, thus diverting the patient from obtaining established medical therapy. Exploitation may delude the public into believing that acupuncture is good for whatever ails you. Acupuncture may indeed have considerable merit and may eventually find an important role in American medicine. That role can only be determined by objective evaluation over a period of years.

The American Society of Anesthesiologists, therefore, was neither accepting nor rejecting the use of acupuncture, but instead it was simply arguing for rigorous testing. These levelheaded experts were not interested in anecdotes, but rather they wanted 'objective evaluation' with large numbers of patients. In other words, they wanted to see acupuncture submitted to the sort of clinical trials discussed in Chapter 1, which had decided the effectiveness of treatments such as bloodletting and lemon juice for scurvy. Perhaps acupuncture would turn out to be as useless as bloodletting, or perhaps it would be as effective as lemons. There was only one way to find out: do proper research.

During the 1970s universities and hospitals across America began submitting acupuncture to clinical trials, all part of a massive effort to test its impact on a variety of ailments. Some of the trials involved just a handful of patients, whereas others involved dozens. Some tracked the impact of acupuncture in the hours immediately following a one-off treatment, while others looked at long-term treatments and monitored the progress of patients over several weeks or even months. The diseases studied ranged from lower back pain to angina, from migraine to arthritis. Despite the wide variety of clinical trials, they broadly followed the principles that had been laid down by James Lind: take patients with a particular condition, randomly assign them either to an acupuncture group or to a control group, and see if those receiving acupuncture improve more than the control group.

A huge number of trials had been conducted by the end of the decade, so in 1979 the World Health Organization Inter-regional Seminar asked R. H. Bannerman to summarize the evidence for and against acupuncture. His conclusions shocked sceptics and vindicated the Chinese. In *Acupuncture: the WHO view*, Bannerman stated that there were more than twenty conditions which 'lend themselves to acupuncture treatment', including sinusitis, common cold, tonsillitis, bronchitis, asthma, duodenal ulcers, dysentery, constipation, diarrhoea, headache and migraine, frozen shoulder, tennis elbow, sciatica, low back pain and osteoarthritis.

This WHO document, and other similarly positive commentaries, marked a watershed in terms of acupuncture's credibility in the West. Budding practitioners could now sign up to courses with confidence, safe in the knowledge that this was a therapy that genuinely worked. Similarly, the number of patients waiting for treatment began to rise rapidly, as they became increasingly convinced of the power of acupuncture. For example, by 1990 in Europe alone there were 88,000 acupuncturists and over 20 million patients had received treatment. Many of the acupuncturists were independent practitioners, but slowly the therapy was also becoming part of mainstream medicine. This was highlighted by a British Medical Association survey in 2002, which revealed that roughly half of all practising doctors had arranged acupuncture sessions for their patients.

The only remaining mystery seemed to be the mechanism that was making acupuncture so effective. Although Western doctors were now becoming sympathetic to the notion that needling specific points on the body could lead to apparently dramatic changes in a person's health, they were highly sceptical about the existence of meridians or the flow of Ch'i. These concepts have no meaning in terms of biology, chemistry or physics, but rather they are based on ancient tradition. The contrast between Western incredulity and Eastern confidence in Ch'i and meridians can be traced back to the evolution of the two medical traditions, particularly the way in which the subject of anatomy was treated in the two hemispheres.

Chinese medicine emerged from a society that rejected human dissection. Unable to look inside the body, the Chinese developed a largely imaginary model of human anatomy that was based on the world around them. For example, the human body was supposed to have 365 distinct components, but only because there are 365 days in the year. Similarly, it seems likely that the belief in twelve meridians emerged as a parallel to the twelve great rivers of China. In short, the human body was interpreted as a microcosm of the universe, as opposed to understanding it in terms of its own reality.

The Ancient Greeks also had reservations about using corpses for medical research, but many notable physicians were prepared to break with tradition in order to study the human body. For instance, in the third century BC, Herophilus of Alexandria explored the brain and its connection to the nervous system. He also identified the ovaries and the fallopian tubes, and was credited with disproving the bizarre and widely held view that the womb wandered around the female body. In contrast to the Chinese, European scientists gradually developed an acceptance that dissecting the human body was a necessary part of medical research, so there was steady progress towards establishing an accurate picture of our anatomy.

Autopsies were becoming common by the thirteenth century, and public dissections for the purpose of teaching anatomy were taking place across Europe by the end of the fourteenth century. By the mid-sixteenth century, the practice of dissection for teaching anatomy to medical students had become standard, largely thanks to the influence of such leading figures as Vesalius, who is acknowledged to be the founder of modern anatomy. He argued that a doctor could not treat the human body unless he understood its construction, but unfortunately obtaining bodies was still a problem. This forced Vesalius, in 1536, to steal the body of an executed criminal still chained to the gibbet. His aim was to obtain a skeleton for research. Luckily much of the flesh had already rotted away or had been eaten by animals, so much so that the bones were 'held together by the ligaments alone'. In 1543 he published his masterpiece, *De Corporis Fabrica* or *The Construction of the Human Body*.

Early European anatomists realized that even the most elementary discoveries about the human body could lead to profound revelations about how it functions. For instance, in the sixteenth century an anatomist named Hieronymus

Fabricus discovered that veins contain one-way valves along their length, which implies that blood flows in only one direction. William Harvey used this information to argue in favour of blood circulating around the body, which in turn ultimately led to a clear understanding of how oxygen, nutrients and disease spread through the human body. Today, modern medicine continues to develop by ever-closer examination of human anatomy, with increasingly powerful microscopes for seeing and with ever finer instruments for dissecting. Moreover, today we can gain insights into a living dynamic body, thanks to endoscopes, X-rays, MRI scans, CAT scans and ultrasound – and yet scientists are still unable to find a shred of evidence to support the existence of meridians or Ch'i.

So, if meridians and Ch'i are fictional, then what is the mechanism behind the apparent healing power of acupuncture? Two decades after Nixon's visit to China had re-introduced acupuncture to the West, scientists had to admit that they were baffled over how acupuncture could supposedly treat so many ailments, ranging from sinusitis to gingivitis, from impotence to dysentery. However, when it came to pain relief, there were tentative theories that seemed credible.

The first theory, known as the *gate control theory of pain*, was developed in the early 1960s, a decade before scientists were thinking about acupuncture. A Canadian named Ronald Melzack and an Englishman named Patrick Wall jointly suggested that certain nerve fibres, which conduct impulses from the skin to more central junctions, also have the ability to close a so-called 'gate'. If the gate is closed, then other impulses, perhaps associated with pain, struggle to reach the brain and are less likely to be recognized as pain. Thus relatively minor stimuli might suppress major pain from other sources by shutting the gate before the troubling pain impulse can reach the brain. The gate control theory of pain has become widely accepted as an explanation of why, for example, rubbing a painful limb is soothing. Could gate control, however, explain the effects of acupuncture? Many acupuncturists in the West argued that the sensation caused by an acupuncture needle was capable of shutting gates and blocking major pain, but sceptics pointed out that there was no solid evidence to show that this was the case. The gate control theory of pain was valid in other situations, but acupuncture's ability to exploit it was unproven.

The second theory for explaining the power of acupuncture is based on the existence of chemicals called *opioids*, which act as powerful, natural painkillers. The most important opioids are known as endorphins. Some studies have indeed shown that acupuncture somehow stimulates the release of these chemicals in the brain. Not surprisingly, acupuncturists have welcomed these studies, but again there have been sceptics. They question whether acupuncture can release enough opioids to create any significant pain relief, and they cite other studies that fail to confirm any connection between endorphins and acupuncture.

In short, here were two theories that could potentially explain the powers of acupuncture, but as yet they were both too tentative to convince the medical establishment. So instead of accepting either theory, scientists urged further research. Meanwhile, they also began to propose a separate explanation to account for the pain relief provided by acupuncture. In fact, if correct, this third theory could potentially explain all its supposed benefits, not just pain relief. Unfortunately for acupuncturists, this third theory attributed the impacts of acupuncture to the *placebo effect*, a medical phenomenon with a long and controversial history.

In one sense, any form of treatment that relies heavily on the placebo effect is fraudulent. Indeed, many bogus therapies from the nineteenth century had turned out to be nothing more than placebo-based treatments. In the next section we will explore the placebo effect in detail and see how it might relate to acupuncture. If the placebo effect can successfully explain the apparent benefits of acupuncture, then 2,000 years of Chinese medical expertise would evaporate. If not, then the medical establishment would be forced to take acupuncture seriously.

The power of placebo

The first medical patent issued under the Constitution of the United States was awarded in 1796 to a physician named Elisha Perkins, who had invented a pair of metal rods which he claimed could extract pains from patients. These *tractors*, as he dubbed them, were not inserted into the patient, but were merely brushed over the painful area for several minutes, during which time they would 'draw off the noxious electrical fluid that lay at the root of suffering'. Luigi Galvani had recently shown that the nerves of living organisms responded to 'animal electricity', so Perkins' tractors were part of a growing fad for healthcare based on the principles of electricity.

As well as providing electrotherapeutic cures for all sorts of pains, Perkins claimed that his tractors could also deal with rheumatism, gout, numbness and muscle weakness. He soon boasted of 5,000 satisfied patients and his reputation was buoyed by the support of several medical schools and high-profile figures such as George Washington, who had himself invested in a pair of tractors. The idea was then exported to Europe when Perkins' son, Benjamin, emigrated to London, where he published *The Influence of Metallic Tractors on the Human Body*. Both father and son made fortunes from their devices – as well as charging their own patients high fees for tractor therapy sessions, they also sold tractors to other physicians for the cost of 5 guineas each. They claimed that the tractors were so expensive because they were made of an exotic metal alloy, and this alloy was supposedly crucial to their healing ability.

However, John Haygarth, a retired British physician, became suspicious about the miraculous powers of the tractors. He lived in Bath, then a popular health resort for the aristocracy, and he was continually hearing about cures attributed to Perkins' tractors, which were all the rage. He accepted that patients treated with Perkins' tractors were indeed feeling better, but he speculated that the devices were essentially fake and that their impact was on the mind, not the body. In other words, credulous patients might be merely convincing themselves that they felt better, because they had faith in the much-hyped and expensive Perkins' tractors. In order to test his theory he made a suggestion in a letter to a colleague:

> Let their merit be impartially investigated, in order to support their fame, if it be well-founded, or to correct the public opinion, if merely formed upon delusion . . . Prepare a pair of false Tractors, exactly to resemble the true Tractors. Let the secret be kept inviolable, not only from the patient but also from any other person. Let the efficacy of both be impartially tried and the reports of the effects produced by the true and false Tractors be fully given in the words of the patients.

Haygarth was suggesting that patients be treated with tractors made from Perkins' special alloy and with fake tractors made of ordinary materials to see if there was any difference in outcome. The results of the trial, which was conducted in 1799 at Bath's Mineral Water Hospital and Bristol Infirmary, were exactly as Haygarth had suspected – patients reported precisely the same benefits whether they were being treated with real or fake tractors. Some of the fake, yet effective, tractors were made of bone, slate and even painted tobacco pipes. None of these materials could conduct electricity, so the entire basis of Perkins' tractors was undermined. Instead Haygarth proposed a new explanation for their apparent effectiveness, namely that 'powerful influences upon diseases is produced by mere imagination'.

Haygarth argued that if a doctor could persuade a patient that a treatment would work, then this persuasion alone could cause an improvement in the patient's condition – or it could at least convince the patient that there had been such an improvement. In one particular case, Haygarth used tractors to treat a woman with a locked elbow joint. Afterwards she claimed that her mobility had increased. In fact, close observation showed that the elbow was still locked and that the lady was compensating by increasing the twisting of her shoulder and wrist. In 1800 Haygarth published *Of the Imagination as a Cause and as a Cure of Disorders of the Body*, in which he argued that Perkins' tractors were no more than quackery and that any benefit to the patient was psychological – medicine had started its investigation into what we today would call the *placebo effect*.

The word *placebo* is Latin for 'I will please', and it was used by writers such as Chaucer to describe insincere expressions that nevertheless can be consoling: 'Flatterers are the devil's chaplains that continually sing placebo.' It was not until 1832 that *placebo* took on its specific medical meaning, namely an insincere or ineffective treatment that can nevertheless be consoling.

Importantly, Haygarth realized that the placebo effect is not restricted to entirely fake treatments, and he argued that it also has a role to play in the impact of genuine medicines. For example, although a patient will derive benefit from taking aspirin largely due to the pill's biochemical effects, there can also be an added bonus benefit due to the placebo effect, which is a result of the patient's confidence in the aspirin itself or confidence in the physician who prescribes it. In other words, a genuine medicine offers a benefit that is largely due to the medicine itself and partly due to the placebo effect, whereas a fake medicine offers a benefit that is entirely due to the placebo effect.

As the placebo effect arises out of the patient's confidence in the treatment, Haygarth wondered about the factors that would increase that confidence and thereby maximize the power of the placebo. He concluded that, among other things, the doctor's reputation, the cost of the treatment and its novelty could all boost the placebo effect. Many physicians throughout history have been quick to hype their reputations, link high cost with medical potency and emphasize the novelty of their cures, so perhaps they were already aware of the placebo effect. In fact, prior to Haygarth's experiments, it seems certain that doctors had been secretly exploiting it for centuries. Nevertheless, Haygarth deserves credit for being the first to write about the placebo effect and bringing it out into the open.

Interest in the placebo effect grew over the course of the nineteenth century, but it was only in the 1940s that an American anaesthetist named Henry Beecher established a rigorous programme of research into its potential. Beecher's own interest in the placebo effect was aroused towards the end of the Second World War, when a lack of morphine at a military field hospital forced him to try an extraordinary experiment. Rather than treating a wounded soldier without morphine, he injected saline into the patient and suggested to the soldier that he was receiving a powerful painkiller. To Beecher's surprise, the patient relaxed immediately and showed no signs of pain, distress or shock. Moreover, when morphine supplies ran low again, the sly doctor discovered that he could repeatedly play this trick on patients. Extraordinarily, it seemed that the placebo effect could subdue even the most severe pains. After the war, Beecher established a major programme of research at Harvard Medical School, which subsequently inspired hundreds of other scientists around the world to explore the miraculous power of placebos.

As the twentieth century progressed, research into placebo responses threw up some rather shocking results. In particular, it soon became clear that some well-established treatments benefited patients largely because of the placebo effect. For example, in 1986 a study was conducted with patients who had undergone tooth extraction, and who then had their jaw massaged by an applicator generating ultrasound. These sound waves, whose frequency is too high to be heard, could apparently reduce post-operative swelling and pain. Unknown to the patients or the therapists, the researchers tampered with the apparatus so that there was no ultrasound during half of the sessions. Because nobody can hear ultrasound, those patients not receiving ultrasound did not suspect that anything was wrong. Astonishingly, patients described similar amounts of pain relief regardless of whether the ultrasound was on or off. It seemed that the effect of the ultrasound treatment was wholly or largely due to the placebo effect and had little to do with whether the equipment was working. Thinking back to Haygarth's criteria for a good placebo, we can see that the ultrasound equipment fits the bill – dentists had promoted it as effective, it looked expensive and it was novel.

An even more startling example relates to an operation known as internal mammary ligation, which was used to relieve the pain of angina. The pain is caused by a lack of oxygen, which itself is caused by insufficient blood running through the narrowed coronary arteries. The surgery in question was supposed to tackle the problem by blocking the internal mammary artery in order to force more blood into the coronary arteries. Thousands of patients underwent the operation and afterwards stated that they suffered less pain and could endure higher levels of exercise. However, some cardiologists became sceptical, because autopsies on patients who eventually died revealed no signs of any extra blood flow through the remaining coronary arteries. If there was no significant improvement in blood flow, then what was causing the patients to improve? Could the relief of symptoms be due simply to the placebo effect? To find out, a cardiologist named Leonard Cobb conducted a trial in the late 1950s that today seems shocking.

Patients with angina were divided into two groups, one of which underwent the usual internal mammary ligation, while the other group received sham surgery; this means that an incision was made in the skin and the arteries were exposed, but no further surgery was conducted. It is important to point out that patients had no idea whether they had undergone the real or sham surgery, as the superficial scar was the same for both. Afterwards, roughly three-quarters of the patients in *both* groups reported significantly lower levels of pain, accompanied by higher exercise tolerance. Incredibly, because both real and sham operations were equally successful, then the surgery itself must have been ineffective and any benefit to the patient must have been induced by a powerful placebo effect. Indeed, the placebo effect was so great that it allowed patients in both groups to reduce their intake of medication.

Although this suggests that the placebo effect is a force for good, it is important to remember that it can have negative consequences. For example, imagine a patient who feels better because of a placebo response to an otherwise ineffective treatment – the underlying problem would still persist, and further treatment might still be necessary, but the temporarily improved patient is less likely to seek that treatment. In the case of mammary ligation, the underlying problem of narrowed arteries and lack of oxygen supply still existed in patients, so they were probably lulled into a false sense of security.

So far, it would be easy to think that the placebo effect is restricted to reducing the experience of pain, perhaps by increasing the patient's pain threshold through placebo-induced will power. Such a view would underestimate the power and scope of the placebo effect, which works for a wide range of conditions, including insomnia, nausea and depression. In fact, scientists have observed real physiological changes in the body, suggesting that the placebo effect goes far beyond the patient's mind by also impacting directly on physiology.

Because the placebo effect can be so dramatic, scientists have been keen to understand exactly how it influences a patient's health. One theory is that it might be related to unconscious *conditioning*, otherwise known as the Pavlovian response, named after Ivan Pavlov. In the 1890s Pavlov noticed that dogs not only salivated at the sight of food, but also at the sight of the person who usually fed them. He considered that salivating at the sight of food was a natural or unconditioned response, but that salivating at the sight of the feeder was an unnatural or conditioned response, which existed only because the dog had come to associate the sight of the person who fed it with the provision of food. Pavlov wondered if he could create other conditioned responses, such as ringing a bell prior to the provision of food. Sure enough, after a while the conditioned dogs would salivate at the sound of the bell alone. The importance of this work is best reflected by the fact that Pavlov went on to win the Nobel Prize for Medicine in 1904.

Whilst such conditioned salivation might seem very different from the placebo effect on health, work by other Russian scientists then went on to show that even an animal's immune response could be conditioned. Researchers worked with guinea pigs, which were known to develop a rash when injected with a certain mildly toxic substance. To see if the rash could be initiated through conditioning, they began lightly scratching the guinea pigs prior to giving an injection. Sure enough, they later discovered that merely scratching the skin and *not* giving the injection could stimulate the same redness and swelling. This was extraordinary – the guinea pig responded to scratching as if it were being injected with the toxin, simply because it had been conditioned to associate strongly the scratching with the consequences of the injection.

So, if the placebo effect in humans is also a conditioned response, then the explanation for its effectiveness would be that a patient simply associates getting better with, for example, seeing a doctor or taking a pill. After all, ever since childhood a patient will have visited a doctor, received a pill and then felt better. Hence, if a doctor prescribes a pill containing no active ingredient, a so-called sugar pill, then the patient might still experience a benefit due to conditioning.

Another explanation for the placebo effect is called the *expectation theory*. This theory holds that if we expect to benefit from a treatment, then we are more likely to do so. Whereas conditioning would exploit our unconscious minds to provoke a placebo response, the expectation theory suggests that our conscious mind might also be playing a role. The expectation theory is supported by a host of data from many lines of research, but it is still poorly understood. One possibility is that our expectations are somehow interacting with our body's so-called *acute phase response*.

The acute phase response covers a range of bodily reactions, such as pain, swelling, fever, lethargy and loss of appetite. In short, the acute phase response is the umbrella term used to describe the body's emergency defensive response to being injured. For instance, the reason that we experience pain is that our body is telling us that we have suffered an injury, and that we need to protect and nurture that part of the body. The experience of swelling is also for our own good, because it indicates an increased blood flow to the injured region, which will accelerate healing. The increased body temperature associated with fever will help kill invading bacteria and provide ideal conditions for our own immune cells. Similarly, lethargy aids recovery by encouraging us to get much-needed rest, and a loss of appetite encourages even more rest because we have suppressed the need to hunt for food. It is interesting to note that the placebo effect is particularly good at addressing issues such as pain, swelling, fever, lethargy and loss of appetite, so perhaps the placebo effect is partly the consequence of an innate ability to block the acute phase response at a fundamental level, possibly by the power of expectation.

The placebo effect may be linked to either conditioning or expectation or both, and there may be other even more important mechanisms that have yet to be identified or fully appreciated. While scientists strive to establish the scientific basis of the placebo effect, they have already been able to ascertain, by building on Haygarth's early work, how to maximize it. It is known, for instance, that a drug administered by injection has a bigger placebo effect than the same drug taken in pill form, and that taking two pills provokes a greater placebo response than taking just one. More surprisingly, green pills have the strongest placebo effect on relieving anxiety, whereas yellow pills work best for depression. Moreover, a pill's placebo effect is increased if it is given by a doctor wearing a white coat, but it is reduced if it is administered by a doctor wearing a T-shirt, and it is even less effective if given by a nurse. Large tablets offer a stronger placebo effect than small tablets . . . unless the tablets are very, very small. Not surprisingly, tablets in fancy branded packaging give a bigger placebo effect than those in plain packets.

Of course, all of the above statements refer to the average patient, because the actual placebo effect for a particular patient depends entirely on the belief system and personal experiences of that individual. This variability of placebo effect among patients, and its potentially powerful influence on recovery, means that it can be a highly misleading factor when it comes to assessing the true efficacy of a treatment. In fact, the placebo effect is so unpredictable that it could easily distort the results of a clinical trial. Therefore, in order to test the true value of acupuncture (and medicines in general), researchers somehow needed to take into account the quirky, erratic and sometimes strong influence of the placebo effect. They would succeed in this endeavour by developing an almost foolproof form of the clinical trial.

The blind leading the double-blind

The simplest form of clinical trial involves a group of patients who receive a new treatment being compared against a group of similar patients who receive no treatment. Ideally there should be a large number of patients in each group and they should be randomly assigned. If the treated group then shows more signs of recovery on average than the untreated control group, then the new treatment is having a real impact . . . or is it?

We must now also consider the possibility that a treatment might have appeared to be effective in the trial, but only because of the placebo effect. In other words, the group of patients being actively treated might expect to recover simply because they are receiving some form of medical intervention, thus stimulating a beneficial placebo response. Hence, the straightforward trial design can produce misleading results, because even a useless treatment can give positive results in such a trial. So the question arises: how do we design a clinical trial that takes into account the confusion caused by the placebo effect?

A solution can be traced back to eighteenth-century France and the extraordinary claims of Franz Mesmer. Whilst Mesmer is nowadays associated with hypnotism (or mesmerism), in his own lifetime he was most famous for promoting the health benefits of magnetism. He argued that he could cure patients of many illnesses by manipulating their 'animal magnetism', and one of the ways of doing this was to give them magnetically treated water. The remedy was very dramatic, because sometimes the supposedly magnetized water could induce fits or fainting as part of the alleged healing process. Critics, however, doubted that water could be magnetized and they were also dubious about the notion that magnetism could affect human health. They suspected that the reactions of Mesmer's patients were purely based on their faith in his claims. In modern parlance, critics were suggesting that Mesmer's remedies were exploiting the placebo effect.

In 1785, Louis XVI convened a Royal Commission to test Mesmer's claims. This Commission, which included Benjamin Franklin, conducted a series of experiments in which one mesmerized glass of water was placed among four glasses of plain water – all five glasses looked identical. Unaware which glass was which, volunteers then randomly picked one glass of water and drank it. In one case, a female patient tasted her glass and immediately fainted, but it was then revealed that she had drunk only plain water. It seemed obvious that the fainting woman thought that she was drinking magnetized water, she knew what was supposed to happen when people drank such water, and her body responded appropriately.

After all the experiments had been completed, the Royal Commission could see that patients had responded in a similar way regardless of whether the water was plain or magnetized. Therefore, they concluded that magnetized water was the same as plain water, which meant that the term magnetized water did not really mean anything. Moreover, the Commission stated that the effect of supposedly magnetized water was due to the expectation of patients; today we would say that it was due to the placebo effect. In short, the Commission accused Mesmer's therapy of being fraudulent.

The Royal Commission did not, however, speculate about the widespread effects of placebo throughout medicine, which is why Haygarth's research on tractors fourteen years later is credited with formally recognizing the role of the placebo effect in medical practice. On the other hand, the Royal Commission did make a major contribution to the history of medicine, because it had designed a new type of clinical trial. The key breakthrough in the Royal Commission's experiment was that the patients were unaware of whether or not they were receiving the real or fake treatment, because the glasses of mesmerized water and plain water appeared to be identical. The patients were said to be *blind*.

The concept of blinding can be applied to entire trials, which are known as *blinded clinical trials*. For example, if a new pill is being tested then it is given to all the patients in the treatment group, while a pill that looks the same but without any active ingredient is given to the control group. Importantly, patients have no idea if they are in the treatment or control group, so they remain blind as to whether or not they are being treated. It is quite possible that both groups will show signs of improvement if both respond to the placebo effect caused by the possibility of receiving the real pill. However, the treatment group should show greater signs of improvement than the control group if the real pill has a genuine effect beyond placebo.

In a blinded trial, it is crucial that both the control group and the treatment group are treated in similar ways, because any variation can potentially affect the recovery of patients and bias the results of the trial. Therefore, as well as receiving pills that look the same, patients in both groups should also be treated in the same location, be given the same level of attention and so on. All these factors can contribute to so-called *non-specific effects* – namely effects resulting from the context of the treatment process, but which are not directly due to the treatment itself. Non-specific effects is the umbrella term that also covers the placebo effect.

It is even necessary to monitor patients in both groups in exactly the same way, because it has been shown that the act of close monitoring can lead to a generally positive change in a person's health or performance. This is known as the Hawthorne effect, a term that was coined after researchers visited the Hawthorne Plant in Illinois, part of the Western Electric Company. The researchers wanted to see how the working environment affected the plant's output, so between 1927 and 1932 they increased artificial illumination and then reduced it, they increased room temperature and then reduced it, and so on. The researchers were amazed to find that any change seemed to cause an improvement. This was partly because workers expected that the changes were supposed to bring about improvements, and partly because they knew they were being monitored by experts with clipboards. It is difficult to remove the Hawthorne effect in any medical trial, but at least the effect should be the same for both the treatment group and the control group so that a fair comparison can be made.

Creating identical conditions for the control and the treatment groups effectively blinds the patients to whether or not they are receiving the treatment or the placebo. Yet it is also important to blind whoever is administering the treatment or the placebo. In other words, even the doctors treating the patients should not be aware of whether they are giving a sugar pill or an active pill. This is because a doctor's demeanour, enthusiasm and tone of voice can all be affected by knowing that he or she is administering a placebo, which means that the doctor might unconsciously give hints to patients that the medicine is merely a placebo. Such leaking of information, of course, can jeopardize the blinding of the patient and the overall reliability of the clinical trial. The consequence would be that patients in the placebo control group would suspect that they were receiving a placebo and would then fail to exhibit a placebo response. Perversely, patients receiving the real treatment would have no such qualms and would exhibit a placebo response. Hence, the trial would be unfair.

If, however, both the patient and the doctor are unaware of whether a placebo or a supposedly active treatment is being administered, then the trial results cannot be influenced by the expectation of either. This type of truly fair trial is said to be *double-blind*. Including some of the points made in Chapter 1, we can now see that a well-conducted trial ideally requires several key features:

1. A comparison between a control group and a group receiving the treatment being tested.
2. A sufficiently large number of patients in each group.
3. Random assignment of patients to each group.
4. The administering of a placebo to the control group.
5. Identical conditions for the control and treatment groups.
6. Blinding patients so that they are unaware to which group they belong.
7. Blinding doctors so that they are unaware whether they are giving a real or a placebo treatment to each patient.

A trial that includes all these features is known as a randomized, placebo-controlled, double-blind clinical trial, and it is considered to be the highest possible standard of medical testing. Nowadays, the various national bodies responsible for authorizing new treatments will usually make their decisions based on the results obtained from such studies.

Sometimes, however, it is necessary to conduct trials that are closely related to this format, but which do not involve a placebo. For example, imagine that scientists want to test a new drug for a condition that is already treated with a partly effective existing drug. Point 3 indicates that the control group receives only a placebo, but this would be unethical if it deprived patients of the partly effective drug. In this situation, the control group would receive the existing drug and the outcome would be compared against the other group receiving the new drug – the trial would not be placebo-controlled, but there would still be a control, namely the existing drug. Such a trial should still adhere to all the other requirements, such as randomization and double-blinding.

These sorts of clinical trials are invaluable when conducting medical research. Although the results from other types of trial and other evidence might be considered, they are generally deemed to be less convincing when it comes to the key question: is a treatment effective for a particular condition?

Returning to acupuncture, we can re-examine the clinical trials of the 1970s and 1980s – were these trials of high quality and were they properly blinded, or is it possible that the reported benefits of acupuncture were due merely to the placebo effect?

A good example of the type of acupuncture trial that took place during this period was one conducted in 1982 by Dr Richard Coan and his team, who wanted to examine whether or not acupuncture was effective for neck pain. His treatment group consisted of fifteen patients who received acupuncture, while his control group consisted of another fifteen patients who remained on a waiting list. The results would have seemed unequivocal to fans of acupuncture, because 80 per cent of patients in the acupuncture group reported an improvement, compared to only 13 per cent of the control group. The extent of the pain relief in the acupuncture group was so great that they halved their intake of painkillers, whereas the control group reduced their intake of pills by only one tenth.

Comparing the acupuncture group against the control group shows that the improvement due to acupuncture is much greater than can be explained by any natural recovery. However, was the benefit from acupuncture due to psychological or physiological factors or a mix of the two? Did the acupuncture trigger a genuine healing mechanism, or did it merely stimulate a placebo response? The latter possibility has to be treated seriously, because acupuncture has many of the attributes that would make it an ideal placebo treatment: needles, mild pain, the slightly invasive nature, exoticism, a basis in ancient wisdom and fantastic press coverage.

So Dr Coan's clinical trial, along with many of the others conducted in the 1970s and 1980s, suffered from the problem that they could not determine whether acupuncture was offering a real benefit or merely a placebo benefit. The ideal way to find out whether acupuncture was genuinely effective would have been to give a placebo to the control group, something that seemed identical to acupuncture but which was totally inert. Unfortunately, finding such a placebo proved difficult – how can you create a therapy that appears to be acupuncture but which is not actually acupuncture? How do you blind patients to whether or not they are receiving acupuncture?

Placebo control groups are easy to arrange in the context of conventional drug trials, because the treatment group can, say, receive a pill with the active ingredient and the placebo control group can receive an identical-looking pill without the active ingredient. Or the treatment group can receive an injection of the active drug and the placebo control group can receive an injection of saline. Unfortunately, there was no similarly obvious placebo replacement for acupuncture.

Gradually, however, researchers began to realize that there were two ways of making patients believe that they were receiving real acupuncture, when they were in fact receiving fake acupuncture. One option was to needle patients to only a minimal depth, as opposed to the centimetre or more that most practitioners would use. The purpose of this superficial needling was that it seemed like the real thing to patients who had not previously experienced genuine acupuncture, but according to the Chinese theory it should have no medical benefit because the needles would not reach the meridian. Therefore researchers proposed studies in which a control group would receive superficial needling, while a treatment group would receive real acupuncture. Both groups would receive similar levels of placebo benefit, but if real acupuncture has a real physiological effect then the treatment group should receive a significant extra benefit beyond that received by the control group.

Another attempt at placebo acupuncture involved needling at points that are not acupuncture points. Such points traditionally have nothing to do with a patient's health. This misplaced needling would seem like genuine acupuncture to new patients, but according to the Chinese theory misplaced needling should have no medical benefit because it would miss the meridians. Hence, some trials were planned in which the control group would receive misplaced needling and the treatment group would receive genuine acupuncture. Both groups would receive the benefit of the placebo effect, but any extra improvement in the treatment group could then be attributed to acupuncture.

These two forms of placebo acupuncture, misplaced and superficial, are often termed sham needling. During the 1990s, sceptics pushed for a major reassessment of acupuncture, this time with placebo-controlled clinical trials involving sham needling. For many acupuncturists, such research was redundant because they had seen how their own patients had responded so positively. They argued that the evidence in favour of their treatment was already compelling. When critics continued to demand placebo-controlled trials, the acupuncturists accused them of clutching at straws and of being prejudiced against alternative medicine. Nevertheless, those medical researchers who believed in the authority of the placebo-controlled trial refused to back down. They continued to voice their doubt and argued that acupuncture would remain a dubious therapy until it had proved itself in high-quality clinical trials.

Those demanding proper acupuncture trials eventually had their wish granted when major funding enabled dozens of placebo-controlled clinical trials to take place in Europe and America throughout the 1990s. Each trial was to be conducted rigorously in the hope that the results would shed new light on who was right and who was wrong. Was acupuncture a miracle medicine that could treat everything from colour blindness to whooping cough, or was it nothing more than a placebo?

Acupuncture on trial

By the end of the twentieth century a new batch of results began to emerge from the latest clinical trials on acupuncture. In general these trials were of higher quality than earlier trials, and some of them examined the impact of acupuncture on conditions that had not previously been tested. With so much new information, the WHO decided that it would take up the challenge of summarizing all the research and presenting some conclusions.

Of course, the WHO had already published a summary document in 1979, which had been very positive about acupuncture's ability to treat more than twenty conditions, but they were keen to revisit the situation in light of the new data that was emerging. The WHO team eventually took into consideration the results from 293 research papers and published their conclusions in 2003 in a report entitled *Acupuncture: Review and analysis of reports on controlled clinical trials.*

The new report assessed the amount and quality of evidence to support the use of acupuncture for a whole series of conditions, and it summarized its conclusions by dividing diseases and disorders into four categories. The first category contained conditions for which there was the most convincing evidence in favour of using acupuncture and the fourth contained conditions for which the evidence was least convincing:

1   Conditions 'for which acupuncture has been proven – through controlled trials – to be an effective treatment' ranged from morning sickness to stroke, and included twenty-eight conditions in total.

2   Conditions 'for which the therapeutic effect of acupuncture has been shown but for which further proof is needed' ranged from abdominal pain to whooping cough, and included sixty-three conditions in total.

3   Conditions 'for which there are only individual controlled trials reporting some therapeutic effects, but for which acupuncture is worth trying because treatment by conventional and other therapies is difficult' ranged from colour blindness to deafness, and included nine conditions in total.

4   Conditions 'for which acupuncture may be tried provided the practitioner has special modern medical knowledge' ranged from coma to convulsions in infants, and included seven conditions in total.

The 2003 WHO report concluded that the benefits of acupuncture were either 'proven' or 'had been shown' in the treatment of ninety-one conditions. It was mildly positive or equivocal about a further sixteen conditions. And the report did not exclude the use of acupuncture for any conditions. The WHO had given acupuncture a ringing endorsement, reinforcing their 1979 report.

It would be natural to assume that this was the final word in the debate over acupuncture, because the WHO is an international authority on medical issues. It would seem that acupuncture had shown itself to be a powerful medical therapy. In fact, the situation is not so clear cut. Regrettably, as we shall see, the 2003 WHO report was shockingly misleading.

The WHO had made two major errors in the way that it had judged the effectiveness of acupuncture. The first error was that they had taken into consideration the results from too many trials. This seems like a perverse criticism, because it is generally considered good to base a conclusion on lots of results from lots of trials involving lots of patients – the more the merrier. If, however, some of the trials have been badly conducted, then those particular results will be misleading and may distort the conclusion. Hence, the sort of overview that the WHO was trying to gain would have been more reliable had it implemented a certain level of quality control, such as including only the most rigorous acupuncture trials. Instead, the WHO had taken into consideration almost every trial ever conducted, because it had set a relatively low quality threshold. Therefore, the final report was heavily influenced by untrustworthy evidence.

The second error was that the WHO had taken into consideration the results of a large number of acupuncture trials originating from China, whereas it would have been better to have excluded them. At first sight, this rejection of Chinese trials might seem unfair and discriminatory, but there is a great deal of suspicion surrounding acupuncture research in China. For example, let's look at acupuncture in the treatment of addiction. Results from Western trials of acupuncture include a mixture of mildly positive, equivocal or negative results, with the overall result being negative on balance. By contrast, Chinese trials examining the same intervention always give positive results. This does not make sense, because the efficacy of acupuncture should not depend on whether it is being offered in the Eastern or Western hemisphere. Therefore, either Eastern researchers or Western researchers must be wrong – as it happens, there are good reasons to believe that the problem lies in the East. The crude reason for blaming Chinese researchers for the discrepancy is that their results are simply too good to be true. This criticism has been confirmed by careful statistical analyses of all the Chinese results, which demonstrate beyond all reasonable doubt that Chinese researchers are guilty of so-called *publication bias*.

Before explaining the meaning of publication bias, it is important to stress that this is not necessarily a form of deliberate fraud, because it is easy to conceive of situations when it can occur due to an unconscious pressure to get a particular result. Imagine a Chinese researcher who conducts an acupuncture trial and achieves a positive result. Acupuncture is a major source of prestige for China, so the researcher quickly and proudly publishes his positive result in a journal. He may even be promoted for his work. A year later he conducts a second similar trial, but on this occasion the result is negative, which is obviously disappointing. The key point is that this second piece of research might never be published for a whole range of possible reasons: maybe the researcher does not see it as a priority, or he thinks that nobody will be interested in reading about a negative result, or he persuades himself that this second trial must have been badly conducted, or he feels that this latest result would offend his peers. Whatever the reason, the researcher ends up having published the positive results of the first trial, while leaving the negative results of the second trial buried in a drawer. This is publication bias.

When this sort of phenomenon is multiplied across China, then we have dozens of published positive trials, and dozens of unpublished negative trials. Therefore, when the WHO conducted a review of the published literature that relied heavily on Chinese research its conclusion was bound to be skewed – such a review could never take into account the unpublished negative trials.

The WHO report was not just biased and misleading; it was also dangerous because it was endorsing acupuncture for a whole range of conditions, some of which were serious, such as coronary heart disease. This begs the question, how and why did the WHO write a report that was so irresponsible?

The WHO has an excellent record when it comes to conventional medicine, but in the area of alternative medicine it seems to prioritize political correctness above truth. In other words, criticism of acupuncture might be perceived as criticism of China, of ancient wisdom and of Eastern culture as a whole. Moreover, usually when expert panels are assembled in order to review scientific research, the protocol is to include experts with informed but diverse opinions. And, crucially, the panel should include critical thinkers who question and challenge any assumptions; otherwise the panel's deliberations are a meaningless waste of time and money. However, the WHO acupuncture panel did not include a single critic of acupuncture. It was quite simply a group of believers who unsurprisingly were less than objective in their assessment. Most worrying of all, the report was drafted and revised by Dr Zhu-Fan Xie, who was Honorary Director of the Institute of Integrated Medicines in Beijing, which fully endorses the use of acupuncture for a range of disorders. It is generally inappropriate for someone with such a strong conflict of interest to be so closely involved in writing a medical review.

If we cannot trust the WHO to summarize adequately the vast number of clinical trials concerning acupuncture, then to whom do we turn? Fortunately, several academics around the world have made up for the WHO's failure by providing their own summaries of the research. Thanks to these groups, we can at long last answer the question that has lingered throughout this chapter – is acupuncture effective?

The Cochrane Collaboration

Doctors are confronted each year with hundreds of new results from clinical trials, which might cover everything from re-testing an existing mainstream treatment to initial testing of a controversial alternative therapy. Often there will be several trials focused on the same treatment for the same ailment, and results can be difficult to interpret and sometimes contradictory. With not enough hours in the day to deal with patients, it would be impractical and nonsensical for doctors to read through each research paper and come to their own conclusions. Instead, they rely heavily on those academics who devote themselves to making sense of all this research, and who publish conclusions that help doctors advise patients about the best form of treatment.

Perhaps the most famous and respected authority in this field is the Cochrane Collaboration, a global network of experts coordinated via its headquarters in Oxford. Firmly adhering to the principles of evidence-based medicine, the Cochrane Collaboration sets itself the goal of examining clinical trials and other medical research in order to offer digestible conclusions about which treatments are genuinely effective for which conditions. Before revealing the Cochrane Collaboration's findings on acupuncture, we will first briefly look at its origins and how it came to be held in such high regard. In this way, by establishing the Cochrane Collaboration's reputation, we hope that you will accept their conclusions about acupuncture in due course.

The Cochrane Collaboration is named after Archie Cochrane, a Scotsman who abandoned his medical studies at University College Hospital, London, in 1936 to serve in the Spanish Civil War as part of a Field Ambulance Unit. Then in the Second World War he joined the Royal Army Medical Corps as a captain and served in Egypt, but he was captured in 1941 and spent the rest of the war providing medical help to fellow prisoners. This was when

Archie Cochrane

he first became aware of the importance of evidence-based medicine. He later wrote that the prison authorities would encourage him by claiming that he was at liberty to decide how to treat his patients: 'I had considerable freedom of choice of therapy: my trouble was that I did not know which to use and when. I would gladly have sacrificed my freedom for a little knowledge.' In order to arm himself with more knowledge he conducted his own trials among his fellow prisoners – he earned their support by telling them about James Lind and the role of clinical trials in working out the best treatment for patients with scurvy.

Whilst Cochrane was clearly a fervent advocate of the scientific method and clinical trials, it is important to note that he also realized the medical value of human compassion, as demonstrated by numerous events throughout his life. One of the most poignant examples occurred during his time as a prisoner of war at Elsterhorst, Germany, when he found himself in the hopeless position of treating a Soviet prisoner who was 'moribund and screaming'. All Cochrane could offer was aspirin. As he later recalled:

> I finally instinctively sat down on the bed and took him in my arms, and the screaming stopped almost at once. He died peacefully in my arms a few hours later. It was not the pleurisy that caused the screaming but loneliness. It was a wonderful education about the care of the dying.

After the war, Cochrane went on to have a distinguished career in medical research. This included studying pneumoconiosis in the coal miners of South Wales and becoming Professor of Tuberculosis and Chest Diseases at the Welsh National School of Medicine in 1960. As his career progressed, he became even more passionate about the value of evidence-based medicine and the need to inform doctors about the most effective medicines. At the same time, he realized that doctors struggled to make sense of all the results from all the clinical trials that were being conducted around the world. Hence Cochrane argued that medical progress would be best served if an organization could be established with the responsibility of drawing clear-cut conclusions from the myriad research projects. In 1979 he wrote, 'It is surely a great criticism of our profession that we have not organised a critical summary, by speciality or subspeciality, adapted periodically, of all relevant randomised controlled trials.'

The key phrase in Cochrane's statement was 'a critical summary', which implied that whoever was doing the summary ought to assess critically the value of each trial in order to determine to what extent it should contribute to the final conclusion about whether a particular therapy is effective for a particular condition. In other words, a carefully conducted trial with lots of patients should be taken seriously; a less carefully conducted trial with just a few patients should carry less weight; and a poorly conducted trial should be ignored completely. This type of approach would become known as a *systematic review*. It is a rigorous scientific evaluation of the clinical trials relating to a particular treatment, as opposed to the sort of reports that the WHO was publishing on acupuncture, which were little more than casual uncritical overviews.

An evidence-based approach to medicine, as previously discussed, means looking at the scientific evidence from clinical trials and other sources in order to decide best medical practice. The systematic review is often the final stage of evidence-based medicine, whereby a conclusion is drawn from all the available evidence. Archie Cochrane died in 1988, by which time the ideas of evidence-based medicine and systematic reviews had taken hold in medicine, but it was not until 1993 that his vision was fully realized with the establishment of the Cochrane Collaboration. Today it consists of twelve centres around the world and over 10,000 health expert volunteers from over ninety countries, who trawl through clinical trials in order 'to help people make well-informed decisions by preparing, maintaining and promoting the accessibility of systematic reviews of the effects of interventions in all areas of health care'.

Having been in existence for over a decade, the Cochrane Collaboration has by now accumulated a library consisting of the results of thousands of trials and has published hundreds of systematic reviews. As well as providing judgements on the effectiveness of pharmaceutical drugs, these systematic reviews evaluate all sorts of other treatments, as well as preventative measures, the value of screening, and the impact of lifestyle and diet on health. In each case, the wholly independent Cochrane Collaboration presents its conclusions about the effectiveness of whatever is being systematically reviewed.

Hopefully this background to the Cochrane Collaboration

has helped to convey its reputation for independence, rigour and quality. This means that we can now look at their systematic reviews of acupuncture and can confidently assume that their conclusions are very likely to be accurate. The Cochrane Collaboration has published several systematic reviews relating to the impact of acupuncture on a variety of conditions, focusing largely on the evidence from placebo-controlled clinical trials.

First, here is the bad news for acupuncturists. The Cochrane reviews suggest that there is no significant evidence to show that acupuncture is an effective treatment for any of the following conditions: smoking addiction, cocaine dependence, induction of labour, Bell's palsy, chronic asthma, stroke rehabilitation, breech presentation, depression, epilepsy, carpal tunnel syndrome, irritable bowel syndrome, schizophrenia, rheumatoid arthritis, insomnia, non-specific back pain, lateral elbow pain, shoulder pain, soft tissue shoulder injury, morning sickness, egg collection, glaucoma, vascular dementia, period pains, whiplash injury and acute stroke. Having examined scores of clinical trials, the Cochrane reviews conclude that any perceived benefit from acupuncture for these conditions is merely a placebo effect. The summaries contain the following sorts of conclusions:

'Acupuncture and related therapies do not appear to help smokers who are trying to quit.'

'There is currently no evidence that auricular acupuncture is effective for the treatment of cocaine dependence.'

'There is insufficient evidence describing the efficacy of acupuncture to induce labour.'

'The current evidence does not support acupuncture as a treatment for epilepsy.'

Also, the Cochrane reviews regularly criticize the quality of the research conducted to date, with comments such as: 'The quality of the included trials was inadequate to allow any conclusion.' Whether the trials were reliable or unreliable, the upshot is the same: despite thousands of years of use in China and decades of scientific research from many countries, there is no sound evidence to support the use of acupuncture for any of the disorders named above.

This is particularly worrying in light of the sort of treatments currently being offered by many acupuncture clinics. For example, by searching for a UK acupuncturist on the web and clicking on the first advert, it was simple to find a central London clinic offering acupuncture for the treatment of all of the following conditions: addictions, anxiety, circulatory problems, depression, diabetes, facial rejuvenation, fatigue, gastrointestinal problems, hay fever, heart problems, high blood pressure, six categories of infertility, insomnia, kidney disorders, liver disease, menopausal problems, menstrual problems, pregnancy care, birth induction, morning sickness, breech presentation, respiratory conditions, rheumatism, sexual problems, sinus problems, skin problems, stress-related illness, urinary problems and weight loss. These conditions fall into one of three categories:

1   Cochrane reviews deem that the evidence from clinical trials does not show acupuncture to be effective.

2   Cochrane reviews conclude that the clinical trials have been so poorly conducted that nothing can be said about the effectiveness of acupuncture with any confidence.

3   The research is so poor and so minimal that the Cochrane Collaboration has not even bothered conducting a systematic review.

Moreover, systematic reviews by other institutions and universities come to exactly the same sort of conclusions arrived at by the Cochrane Collaboration. Despite the fact that there is no reason to believe that it works for any of these conditions, except as a placebo, thousands of clinics in Europe and America are still willing to promote acupuncture for such a wide-ranging list of ailments.

The good news for acupuncturists is that the Cochrane reviews have been more positive about acupuncture's ability to treat other conditions. There have been cautiously optimistic Cochrane reviews on the treatment of pelvic and back pain during pregnancy, low back pain, headaches, postoperative nausea and vomiting, chemotherapy-induced nausea and vomiting, neck disorders and bedwetting. Aside from bedwetting, the only positive conclusions relate to acupuncture in dealing with some types of pain and nausea.

Although these particular Cochrane reviews are the most positive about acupuncture's benefits, it is important to note that their support is only half-hearted. For example, in the case of idiopathic headaches, namely those that occur for no known reason, the review states: 'Overall, the existing evidence supports the value of acupuncture for the treatment of idiopathic headaches. However, the quality and amount of evidence are not fully convincing.'

Because the evidence is only marginally positive and not fully convincing, even in the areas of pain and nausea, researchers have focused their efforts on improving the quality and amount of evidence in order to reach a more concrete conclusion. Indeed, one of the authors of this book, Professor Edzard Ernst, has been part of this effort. Ernst, who leads the Complementary Medicine Research Group at the University of Exeter, became interested in acupuncture when he learned about it at medical school. Since then, he has visited acupuncturists in China, conducted ten of his own clinical trials, published more than forty reviews examining other acupuncture trials, written a book on the subject and currently sits on the editorial board of several acupuncture journals. This demonstrates his commitment to investigating with an open mind the value of this form of treatment, while thinking critically and helping to improve the quality of acupuncture trials.

One of Ernst's most important contributions to improving the quality of trials has been to develop a superior form of sham acupuncture, something even better than misplaced or superficial needling. Figure 1 on page 62 shows how an acupuncture device consists of a very fine needle and a broader upper part that is held by the acupuncturist. Ernst and his colleagues proposed the idea of a telescopic needle – that is, an acupuncture needle that looks as if it penetrates the skin, but which instead retracts into the upper handle part, rather like a theatrical dagger.

Jongbae Park, a Korean PhD student in Ernst's group, went ahead and built a prototype, overcoming various problems along the way. For example, usually an acupuncture needle stays in place because it is embedded in the skin, but the telescopic needle would only appear to penetrate the skin, so how would it stay upright? The solution was to rely on the plastic guide tube, which acupuncturists often use to help position and ease needle insertion. The guide tube is usually removed after insertion, but Park suggested making one end of the tube sticky and leaving it in place so that it could support the needle. Park also designed the telescopic system so that the needle offered some resistance as it retracted into the upper handle. This meant that it would cause some minor sensation during its apparent insertion, which in turn would help convince the patient that this was real acupuncture that was being practised.

When the Exeter group tested these telescopic needles as

part of a placebo acupuncture session, patients were indeed convinced that they were receiving real treatment. They saw the long needle, watched it shorten on impact with the skin, felt a small, localized pain and saw the needle sitting in place for several minutes before being withdrawn. Superficial and misplaced needling were adequate placebos, but an ideal acupuncture placebo should not pierce the skin, which is why this telescopic needling was a superior form of sham therapy. The team was delighted to have developed and validated the first true placebo for acupuncture trials, though their pride was tempered when they discovered that two German research groups at Heidelberg and Hannover Universities had been working on a very similar idea. Great minds were thinking alike.

It has taken several years to design, develop and test the telescopic needle, and it has taken several more years to arrange and conduct clinical trials using it. Now, however, the first results have begun to emerge from what are arguably the highest-quality acupuncture trials ever conducted.

These initial conclusions have generally been disappointing for acupuncturists: they provide no convincing evidence that real acupuncture is significantly more effective than placebo acupuncture in the treatment of chronic tension headache, nausea after chemotherapy, post-operative nausea and migraine prevention. In other words, these latest results contradict some of the more positive conclusions from Cochrane reviews. If these results are repeated in other trials, then it is probable that the Cochrane Collaboration will revise its conclusions and make them less positive. In a way, this is not so surprising. In the past, when trials were poorly conducted, the results for acupuncture seemed positive; but when the trials improved in quality, then the impact of acupuncture seemed to fade away. The more that researchers eliminate bias from their trials, the greater the tendency for results to indicate that acupuncture is little more than a placebo. If researchers were able to conduct perfect trials, and if this trend continues, then it seems likely that the truth is that acupuncture offers negligible benefit.

Unfortunately, it will never be possible to conduct a perfect acupuncture trial, because the ideal trial is double-blind, meaning that neither the patient nor the practitioner knows if real or placebo treatment is being given. In an acupuncture trial, the practitioner will always know if the treatment is real or a placebo. This might seem unimportant, but there is a risk that the practitioner will unconsciously communicate to the patient that a placebo is being administered, perhaps because of the practitioner's body language or tone of voice. It could be that the marginally positive results for acupuncture for pain relief and nausea apparent in some trials are merely due to the slight remaining biases that occur with single blinding. The only hope for minimizing this problem in future is to give clear and strong guidance to practitioners involved in trials to minimize inadvertent communication.

While some scientists have focused on the use of telescopic needles in their trials, German researchers have concentrated on involving larger numbers of patients in order to improve the accuracy of their conclusions. German interest in testing acupuncture dates back to the late 1990s, when the national authorities voiced serious doubts about the entire field. They questioned whether they should continue paying for acupuncture treatment in the light of the lack of reliable evidence. To remedy the situation, Germany's Federal Committee of Physicians and Health Insurers took a dramatic step and decided to initiate eight high-quality acupuncture trials, which would examine four ailments: migraine, tension-type headache, chronic low back pain and knee osteoarthritis. These trials were to involve more patients than any previous acupuncture trial, which is why they became known as mega-trials.

The number of patients in the trials ranged from 200 to over 1,000. Each trial divided its patients into three groups: the first group received no acupuncture, the second group received real acupuncture, and the third (placebo) group received sham acupuncture. In terms of sham acupuncture, the researchers did not employ the new stage-dagger needles, as they had only just been invented and had not yet been properly assessed. Instead, sham acupuncture took the form of misplaced or superficial needling.

Due to their sheer size, these mega-trials have taken many years to conduct. They were completed only recently and the emerging data is still being analysed. Nevertheless, by 2007 the researchers published their initial conclusions from all the mega-trials. They indicate that real acupuncture performs only marginally better than or the same as sham acupuncture. The conclusions typically contain the following sort of statement: 'Acupuncture was no more effective than sham acupuncture in reducing migraine headaches.' Again, the trend continues – as the trials become increasingly rigorous and more reliable, acupuncture increasingly looks as if it is nothing more than a placebo.

Conclusions

The history of acupuncture research has followed a tortuous path over the last three decades, and more research papers will be published in the future, particularly making use of the relatively new telescopic sham needles and with a fuller evaluation of the German mega-trials. However, the research is already fitting together well, with a high level of consistency and agreement. Hence, it seems likely that our current understanding of acupuncture is fairly close to the truth, and we will conclude this chapter with a summary of what we know from the mass of research. The four key outcomes are as follows:

1   The traditional principles of acupuncture are deeply flawed, as there is no evidence at all to demonstrate the existence of Ch'i or meridians.

2   Over the last three decades, a huge number of clinical trials have tested whether or not acupuncture is effective for treating a variety of disorders. Some of these trials have implied that acupuncture is effective. Unfortunately, most of them have been without adequate placebo control groups and of poor quality – the majority of positive trials are therefore unreliable.

3   By focusing on the increasing number of high-quality research papers, reliable conclusions from systematic reviews make it clear that acupuncture does not work for a whole range of conditions, except as a placebo. Hence, if you see acupuncture being advertised by a clinic, then you can assume that it does not really work, except possibly in the treatment of some types of pain and nausea.

4   There are some high-quality trials that support the use of acupuncture for some types of pain and nausea, but there are also high-quality trials that contradict this conclusion. In short, the evidence is neither consistent nor convincing – it is borderline.

These four points also apply to variations of acupuncture, such as acupressure (needles are replaced by pressure applied by fingers or sticks), moxibustion (ground mugwort herb burns above the skin and heats acupuncture points), and forms of acupuncture involving electricity, laser light or sound vibrations. These therapies are based on the same core principles, and it is simply a question of whether the acupuncture points are pricked, pressurized, heated, electrified, illuminated or oscillated. These more exotic forms of acupuncture have been less rigorously tested than conventional acupuncture, but the overall conclusions are similarly disappointing.

In summary, if acupuncture were to be considered in the same way that a new conventional painkilling drug might be tested, then it would have failed to prove itself and would not be allowed into the health market. Nevertheless, acupuncture has grown to become a multi-billion-pound worldwide business that exists largely outside mainstream medicine. Acupuncturists would argue that this industry is legitimate, because there is some evidence that acupuncture works. Critics, on the other hand, would point out that the majority of acupuncturists treat disorders for which there is no respectable evidence whatsoever. And, even in the case of treating pain and nausea, critics would argue that the benefits of acupuncture (if they exist at all) must be relatively small – otherwise these benefits would already have been demonstrated categorically in clinical trials. Moreover, there are conventional painkilling drugs that can achieve levels of pain relief with reasonable reliability, which are vastly cheaper than acupuncture sessions. After all, an acupuncture session costs at least £25 and a full course may run to dozens of sessions.

When medical researchers argue that the evidence seems largely to disprove the benefits of acupuncture, the response from acupuncturists often includes five main criticisms. Although superficially persuasive, these criticisms are based on very weak arguments. We shall address them one by one:

1   Acupuncturists point out that we cannot simply ignore those randomized placebo-controlled clinical trials that indicate that acupuncture works. Of course, such evidence should not be ignored, but it has to be weighed

against the evidence that counters it, and we need to decide which side of the argument is more convincing, much as a jury would do in a legal case. So let us weigh up the evidence. Is acupuncture effective for a wide range of disorders beyond all reasonable doubt? No. Is acupuncture effective for pain and nausea beyond all reasonable doubt? No. Is acupuncture effective for pain and nausea on the balance of probabilities? The jury is still out, but as time has passed and scientific rigour has increased, then the balance of evidence has moved increasingly against acupuncture. For example, as this book goes to print, the results have emerged of a clinical trial involving 640 patients with chronic back pain. According to this piece of research, which was sponsored by the National Institute of Health in America and conducted by Daniel Cherkin, sham acupuncture is just as effective as real acupuncture. This supports the view that acupuncture treatment acts as nothing more than a powerful placebo.

2 Practitioners argue that acupuncture, like many alter native therapies, is an individualized, complex treatment and therefore is not suitable for the sort of large-scale testing that is involved in a trial. This argument is based on the misunderstanding that clinical trials necessarily disregard individualization or complexity. The truth is that such features can be (and often are) incorporated into the design of clinical trials. Furthermore, most conventional medicine is equally complex and individualized, and yet it has progressed thanks to clinical trials. For instance, a doctor will ask a patient about his or her medical history, age, their general health, any recent changes in diet or routine and so on. Having considered all these factors, the doctor will offer a treatment

appropriate to that individual patient – that treatment is likely to have been tested in a randomized clinical trial.

3   Many acupuncturists claim that the underlying philosophy of their therapy is so at odds with conventional science that the clinical trial is inappropriate for testing its efficacy. But this accusation is irrelevant, because clinical trials have nothing to do with philosophy. Instead, clinical trials are solely concerned with establishing whether or not a treatment works.

4   Acupuncturists complain that the clinical trial is inappropriate for alternative therapies because the impact of the treatment is very subtle. But if the effect of acupuncture is so subtle that it cannot be detected, then is it really a worthwhile therapy? The modern clinical trial is a highly sophisticated, flexible and sensitive approach to assessing the efficacy of any treatment and it is the best way of detecting even the most subtle effect. It can measure effects in all sorts of ways, ranging from analysing a patient's blood to asking a patient to assess their own health. Some trials use well-established questionnaires that require patients to report several aspects of their quality of life, such as physical pain, emotional problems and vitality.

5   Finally, some acupuncturists point out that real acupuncture may perform only as well as sham acupuncture, but what if sham acupuncture offers a genuine medical benefit to patients? We have assumed so far that sham acupuncture is inert, except as a placebo, but is it conceivable that superficial and misplaced needling also somehow tap into the body's meridians? If this turns out to be true, then the entire philosophy of

acupuncture falls apart – inserting a needle anywhere to any depth would have a therapeutic benefit, which seems highly unlikely. Also, the development of the telescopic needle sidesteps this question because it does not puncture the skin, so it cannot possibly tap into any meridians. Acupuncturists might counter by arguing that telescopic needles also offer therapeutic benefit because they apply pressure to the skin, but if this were the case then we would also receive benefits from a handshake, a tap on the back or scratching an ear. Alternatively, such pressure on the skin might sometimes detrimentally influence the flow of Ch'i, so such bodily contact might make us ill. This is clearly a ridiculous criticism.

In short, none of these criticisms stands up to proper scrutiny. They are the sort of flimsy arguments that one might expect from practitioners who instinctively want to protect a therapy in which they have both a professional and an emotional investment. Such acupuncturists are unwilling to accept that the clinical trial is undoubtedly the best method available for minimizing bias. Although never perfect, the clinical trial allows us to get as close to the truth as we possibly can.

In fact, it is important to remember that the clinical trial is so effective at minimizing bias that it is also a vital tool in researching conventional medicine. This is a point that was well made by the British Nobel Prize-winning scientist Sir Peter Medawar:

> Exaggerated claims for the efficacy of a medicament are very seldom the consequence of any intention to deceive; they are usually the outcome of a kindly conspiracy in which everybody has the very best intentions. The patient wants to get well, his physician wants to have made him better, and the

pharmaceutical company would have liked to have put it into the physician's power to have made him so. The controlled clinical trial is an attempt to avoid being taken in by this conspiracy of good will.

Although this chapter demonstrates that acupuncture is very likely to be acting as nothing more than a placebo, we cannot end it without raising one issue that might rescue the role of acupuncture within a modern healthcare system. We have already seen that the placebo effect can be a very powerful and positive influence in healthcare, and acupuncture seems to be very good at eliciting a placebo response. Hence, can acupuncturists justify their existence by practising placebo medicine and helping patients with an essentially fake treatment?

For example, we explained that the German mega-trials divided patients into three groups: one received real acupuncture, one received sham acupuncture, and one received no acupuncture at all. In general, the results showed that real acupuncture significantly reduced pain in about half of patients and sham acupuncture delivered roughly the same level of benefit, while the third group of patients showed significantly less improvement. The fact that real and sham acupuncture are roughly as effective as each other implies that real acupuncture merely exploits the placebo effect – but does this matter as long as patients are deriving benefit? In other words, does it matter that the treatment is fake, as long as the benefit is real?

A treatment that relies so heavily on the placebo effect is essentially a bogus treatment, akin to Mesmer's magnetized water and Perkins' tractors. Acupuncture works only because the patients have faith in the treatment, but if the latest research were to be more strongly promoted, then some patients would lose their confidence in acupuncture and the placebo benefits would largely melt away. Some people might therefore argue that there should be a conspiracy of silence so that the mystique and power of acupuncture is maintained, which in turn would mean that patients could continue to benefit from needling. Others might feel that misleading patients is fundamentally wrong and that administering placebo treatments is unethical.

The issue of whether or not placebo therapies are acceptable will be relevant to some other forms of alternative medicine, so this issue will be fully addressed in the final chapter. In the meantime, the main question is: which of the other major alternative therapies are genuinely effective, and which are merely placebos?

3 The Truth About Homeopathy

'Truth is tough. It will not break, like a bubble, at a touch; nay, you may kick it about all day, like a football, and it will be round and full at evening.'

Oliver Wendell Holmes, Sr

Homeopathy
(or Homoeopathy)

A system for treating illness based on the premise that like cures like. The homeopath treats symptoms by administering minute or non-existent doses of a substance which in large amounts produces the same symptoms in healthy individuals. Homeopaths focus on treating patients as individuals and claim to be able to treat virtually any ailment, from colds to heart disease.

IN RECENT DECADES HOMEOPATHY HAS BECOME ONE OF THE fastest-growing forms of alternative medicine, particularly in Europe. The proportion of the French population using homeopathy increased from 16 per cent to 36 per cent between 1982 and 1992, while in Belgium over half the population regularly relies on homeopathic remedies. This rise in demand has encouraged more people to become practitioners – known as homeopaths – and it has even convinced some conventional doctors to study the subject and offer homeopathic treatments. The UK-based Faculty of Homeopathy already has over 1,400 doctors on its register, but the greatest number of practitioners is in India, where there are 300,000 qualified homeopaths, 182 colleges and 300 homeopathic hospitals. And while America has far fewer homeopaths than India, the profits to be made are much greater. Annual sales in the United States increased five-fold from $300 million in 1987 to $1.5 billion in 2000.

With so many practitioners and so much commercial success, it would be reasonable to assume that homeopathy must be effective. After all, why else would millions of people – educated and uneducated, rich and poor, in the East and the West – rely on it?

Yet the medical and scientific establishment has generally viewed homeopathy with a great deal of scepticism, and its remedies have been at the centre of a long-running and often heated debate. This chapter will look at the evidence and reveal whether homeopathy is a medical marvel or whether the critics are correct when they label it a quack medicine.

The origins of homeopathy

Unlike acupuncture, homeopathy's origins are not shrouded in the mists of time, but can be traced back to the work of a German physician called Samuel Hahnemann at the end of the eighteenth century. Having studied medicine in Leipzig, Vienna and Erlangen, Hahnemann earned a reputation as one of Europe's foremost intellectuals. He published widely on both medicine and chemistry, and used his knowledge of English, French, Italian, Greek, Latin, Arabic, Syriac, Chaldaic and Hebrew to translate numerous scholarly treatises.

He seemed set for a distinguished medical career, but during the 1780s he began to question the conventional practices of the day. For instance, he rarely bled his patients, even though his colleagues strongly advocated bloodletting. Moreover, he was an outspoken critic of those responsible for treating the Holy Roman Emperor Leopold of Austria, who was bled four times in the twenty-four hours immediately prior to his death in 1792. According to Hahnemann, Leopold's high fever and abdominal distension did not require such a risky treatment. Of course, we now know that bloodletting is indeed a dangerous intervention. The imperial court physicians, however, responded by calling Hahnemann a murderer for depriving his own patients of what they deemed to be a vital medical procedure.

Hahnemann was a decent man, who combined intelligence with integrity. He gradually realized that his medical colleagues knew very little about how to diagnose their patients accurately, and worse still these doctors knew even less about the impact of their treatments, which meant that they probably did more harm than good. Not surprisingly, Hahnemann eventually felt unable to continue practising this sort of medicine:

My sense of duty would not easily allow me to treat the unknown pathological state of my suffering brethren with these unknown medicines. The thought of becoming in this way a murderer or malefactor towards the life of my fellow human beings was most terrible to me, so terrible and disturbing that I wholly gave up my practice in the first years of my married life and occupied myself solely with chemistry and writing.

In 1790, having moved away from all conventional medicine, Hahnemann was inspired to develop his own revolutionary school of medicine. His first step towards inventing homeopathy took place when he began experimenting on himself with the drug Cinchona, which is derived from the bark of a Peruvian tree. Cinchona contains quinine and was being used successfully in the treatment of malaria, but Hahnemann consumed it when he was healthy, perhaps in the hope that it might act as a general tonic for maintaining good health. To his surprise, however, his health began to deteriorate and he developed the sort of symptoms usually associated with malaria. In other words, here was a substance that was normally used for curing the fevers, shivering and sweating suffered by a malaria patient, which was now apparently generating the same symptoms in a healthy person.

He experimented with other treatments and obtained the same sort of results: substances used to treat particular symptoms in

Samuel Hahnemann

an unhealthy person seemed to generate those same symptoms when given to a healthy person. By reversing the logic, he proposed a universal principle, namely 'that which can produce a set of symptoms in a healthy individual, can treat a sick individual who is manifesting a similar set of symptoms'. In 1796 he published an account of his *Law of Similars*, but so far he had gone only halfway towards inventing homeopathy.

Hahnemann went on to propose that he could improve the effect of his 'like cures like' remedies by diluting them. According to Hahnemann, and for reasons that continue to remain mysterious, diluting a remedy increased its power to cure, while reducing its potential to cause side-effects. His assumption bears some resemblance to the 'hair of the dog that bit you' dictum, inasmuch as a little of what has harmed someone can supposedly undo the harm. The expression has its origins in the belief that a bite from a rabid dog could be treated by placing some of the dog's hairs in the wound, but nowadays 'the hair of the dog' is used to suggest that a small alcoholic drink can cure a hangover.

Moreover, while carrying his remedies on board a horse-drawn carriage, Hahnemann made another breakthrough. He believed that the vigorous shaking of the vehicle had further increased the so-called *potency* of his homeopathic remedies, as a result of which he began to recommend that shaking (or *succussion*) should form part of the dilution process. The combination of dilution and shaking is known as *potentization*.

Over the next few years, Hahnemann identified various homeopathic remedies by conducting experiments known as *provings*, from the German word *prüfen*, meaning to examine or test. This would involve giving daily doses of a homeopathic remedy to several healthy people and then asking them to keep a detailed diary of any symptoms that might emerge over the course of a few weeks. A compilation of their diaries was then used to identify the range of symptoms suffered by a healthy person taking the remedy – Hahnemann then argued that the identical remedy given to a sick patient could relieve those same symptoms.

In 1807 Hahnemann coined the word *Homöopathie*, from the Greek *hómoios* and *pathos*, meaning similar suffering. Then in 1810 he published *Organon der rationellen Heilkunde (Organon of the Medical Art)*, his first major treatise on the subject of homeopathy, which was followed in the next decade by *Materia Medica Pura*, six volumes that detailed the symptoms cured by sixty-seven homeopathic remedies. Hahnemann had given homeopathy a firm foundation, and the way that it is practised has hardly changed over the last two centuries. According to Jay W. Shelton, who has written extensively on the subject, 'Hahnemann and his writings are held in almost religious reverence by most homeopaths.'

The gospel according to Samuel Hahnemann

Hahnemann was adamant that homeopathy was distinct from herbal medicine, and modern homeopaths still maintain a separate identity and refuse to be labelled herbalists. One of the main reasons for this is that homeopathic remedies are not solely based on plants. They can also be based on animal sources, which sometimes means the whole animal (e.g. ground honeybee), and sometimes just animal secretions (e.g. snake poison, wolf milk). Other remedies are based on mineral sources, ranging from salt to gold, while so-called *nosode* sources are based on diseased material or causative agents, such as bacteria, pus, vomit, tumours, faeces and warts. Since Hahnemann's era, homeopaths have also relied upon an additional set of sources labelled *imponderables*, which covers non-material phenomena such as X-rays and magnetic fields.

There is something innately comforting about the idea of

herbal medicines, which conjures up images of leaves, petals and roots. Homeopathic remedies, by contrast, can sound rather disturbing. In the nineteenth century, for instance, a homeopath describes basing a remedy on 'pus from an itch pustule of a young and otherwise healthy Negro, who had been infected [with scabies]'. Other homeopathic remedies require crushing live bedbugs, operating on live eels and injecting a scorpion in its rectum.

Another reason why homeopathy is absolutely distinct from herbal medicine, even if the homeopathic remedy is based on plants, is Hahnemann's emphasis on dilution. If a plant is to be used as the basis of a homeopathic remedy, then the preparation process begins by allowing it to sit in a sealed jar of solvent, which then dissolves some of the plant's molecules. The solvent can be either water or alcohol, but for ease of explanation we will assume that it is water for the remainder of this chapter. After several weeks the solid material is removed – the remaining water with its dissolved ingredients is called the *mother tincture*.

The mother tincture is then diluted, which might involve one part of it being dissolved in nine parts water, thereby diluting it by a factor of ten. This is called a 1X remedy, the X being the Roman numeral for 10. After the dilution, the mixture is vigorously shaken, which completes the potentization process. Taking one part of the 1X remedy, dissolving it in nine parts water and shaking again leads to a 2X remedy. Further dilution and potentization leads to 3X, 4X, 5X and even weaker solutions – remember, Hahnemann believed that weaker solutions led to stronger remedies. Herbal medicine, by contrast, follows the more commonsense rule that more concentrated doses lead to stronger remedies.

The resulting homeopathic solution, whether it is 1X, 10X or even more dilute, can then be directly administered to a patient as a remedy. Alternatively, drops of the solution can be added to an ointment, tablets or some other appropriate form of delivery. For example, one drop might be used to wet a dozen sugar tablets, which would transform them into a dozen homeopathic pills.

At this point, it is important to appreciate the extent of the dilution undergone during the preparation of homeopathic remedies. A 4X remedy, for instance, means that the mother tincture was diluted by a factor of 10 (1X), then again by a factor of 10 (2X), then again by a factor of 10 (3X), and then again by a factor of 10 (4X). This leads to dilution by a factor of 10 x 10 x 10 x 10, which is equal to 10,000. Although this is already a high degree of dilution, homeopathic remedies generally involve even more extreme dilution. Instead of dissolving in factors of 10, homeopathic pharmacists will usually dissolve one part of the mother tincture in 99 parts of water, thereby diluting it by a factor of 100. This is called a 1C remedy, C being the Roman numeral for 100. Repeatedly dissolving by a factor of 100 leads to 2C, 3C, 4C and eventually to ultra-dilute solutions.

For example, homeopathic strengths of 30C are common, which means that the original ingredient has been diluted 30 times by a factor of 100 each time. Therefore, the original substance has been diluted by a total factor of 1,000,000,000,000,000,000,000,000,000,000,000,000,000, 000,000,000,000,000,000,000. This string of noughts might not mean much, but bear in mind that one gram of the mother tincture contains less than 1,000,000,000,000, 000,000,000,000 molecules. As indicated by the number of noughts, the degree of dilution is vastly bigger than the number of molecules in the mother tincture, which means that there are simply not enough molecules to go round. The bottom line is that this level of dilution is so extreme that the resulting solution is unlikely to contain a single molecule of the original ingredient. In fact, the chance of having one

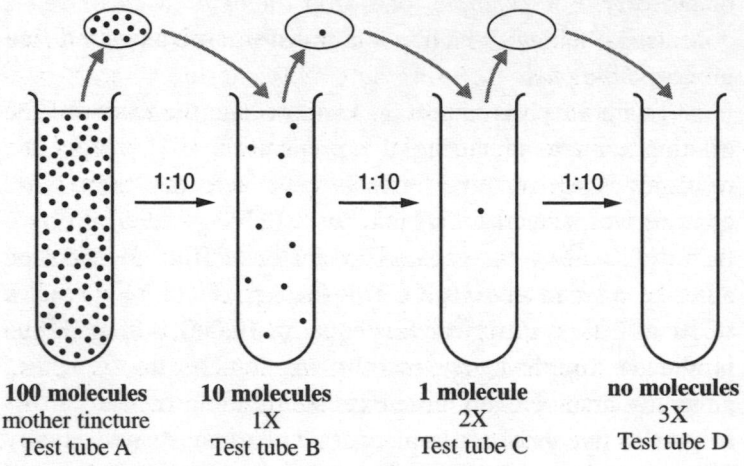

100 molecules
mother tincture
Test tube A

10 molecules
1X
Test tube B

1 molecule
2X
Test tube C

no molecules
3X
Test tube D

Figure 2

Homeopathic remedies are prepared by repeated dilution, with vigorous shaking between stages. Test tube A contains the initial solution, called the mother tincture, which in this case has 100 molecules of the active ingredient. A sample from test tube A is then diluted by a factor of ten (1X), which leads to test tube B, which contains only 10 molecules in a so-called 1X dilution. Next, a sample from test tube B is diluted by a factor of ten again (2X), which leads to test tube C, which contains only 1 molecule. Finally, a sample from test tube C is diluted by a factor of ten for a third time (3X), which leads to test tube D, which is very unlikely to contain any molecules of the active ingredient. Test tube D, devoid of any active ingredient, is then used to make homeopathic remedies. In practice, the number of molecules in the mother tincture will be much greater, but the number of dilutions and the degree of dilution is generally more extreme, so the end result is typically the same – no molecules in the remedy.

molecule of the active ingredient in the final 30C remedy is one in a billion billion billion billion. In other words, a 30C homeopathic remedy is almost certain to contain nothing more than water. This point is graphically explained in Figure 2. Again, this underlines the difference between herbal and homeopathic remedies – herbal remedies will always have at least a small amount of active ingredient, whereas homeopathic remedies usually contain no active ingredient whatsoever.

Materials that will not dissolve in water, such as granite, are ground down and then one part of the resulting powder is mixed with 99 parts lactose (a form of sugar), which is then ground again to create a 1C composition. One part of the resulting powder is mixed with 99 parts lactose to create a 2C composition, and so on. If this process is repeated 30 times, then the resulting powder can be compacted into 30C tablets. Alternatively, at any stage the powder might be dissolved in water and the remedy can be repeatedly diluted as described previously. In either case, the resulting 30C remedy is, again, almost guaranteed to contain no atoms or molecules of the original active granite ingredient.

As if all this was not sufficiently mysterious, some homeo-pathic pharmacies stock 100,000C remedies, which means that the manufacturers are taking 30C remedies, already devoid of any active ingredient, and then diluting them by a factor of 100 another 99,970 times. Because of the time required to make 100,000 dilutions, each one followed by a vigorous shaking, such remedies can cost more than £1,000.

From a scientific perspective, it is impossible to explain how a remedy that is devoid of any active ingredient can have any conceivable effect on any medical condition, apart from the obvious placebo effect. Homeopaths would argue that the remedy has some memory of the original ingredient, which somehow influences the body, but this makes no scientific sense. Nevertheless, homeopaths still claim that their remedies are effective for a whole range of conditions, from temporary problems (coughs, diarrhoea and headaches) to more chronic conditions (arthritis, diabetes and asthma), and from minor ailments (bruises and colds) to more serious conditions (cancer and Parkinson's disease).

Although we have listed various diseases, it is important to point out that Hahnemann and his descendants do not see themselves as treating diseases in a conventional sense. Instead they focus on the individual symptoms and the characteristics of the patients. This is best illustrated by describing how a homeopath typically deals with a case.

The homeopath starts by conducting a detailed interview with the patient, asking about both physical and psychological symptoms. This will result in several pages of notes detailing every symptom, including where each one occurs in the body, when they arise and any activities that affect these symptoms. For example, even if the chief complaint is earache, the notes will include meticulous descriptions of everything from the patient's bunions to any recent constipation. Homeopathy is a highly individualized therapy, so the consultation may even ask about the patient's personality, emotional wellbeing, apparently trivial matters from their past and preferences for food, colours and smells. This whole interview process usually lasts for more than an hour and the outcome is a complete analysis of the patient's symptoms.

As the ultimate goal is to find a homeopathic remedy that best fits all the symptoms that have been described, the next stage is to consult the *Materia Medica*, the encyclopaedia that lists the remedies and what they should be used for. Although Hahnemann identified just a few dozen remedies in his early writings, the homeopath William Boericke included over 600 in his *Materia Medica* in 1901, and today *The Homeopathic Pharmacopoeia of the United States* recognizes over 1,000. Trawling through all these potential remedies is all the more complicated because each remedy treats numerous symptoms and so its entry can run to more than a page. For instance, Figure 3 shows the entry for *Aceticum acidum*, better known as acetic acid or the chemical associated with vinegar.

Ideally the homeopath is looking for the *simillimum*, which is to say the remedy that offers a perfect match with the patient's symptoms. In order to find this optimal remedy, the homeopath might refer to a *repertory*, which is organized according to each symptom followed by the remedies associated with it (as opposed to a *Materia Medica*, which is organized according to each remedy followed by its associated symptoms). Poring through a repertory can still be a major task, so the homeopath will tend to focus on very special and peculiar symptoms to help narrow the search. For example, according to Boericke's *Materia Medica*, 'Face: distortion of mouth, trembling of jaw, facial paralysis; more left side' coupled with 'Stool: bloody, black, and offensive. Gelatinous, yellowish green; semi-fluid, with urinary suppression' means that Cadmium sulphate is the ideal remedy.

Finding the correct remedy is such a complicated and subtle task that a patient who visited different homeopaths and who underwent different interviews would be likely to receive different remedies. In fact, the process of finding the correct remedy can vary so much that it has led to the emergence of distinct schools of homeopathy. For example, *clinical homeopathy* simplifies matters by focusing only on the patient's main symptom and ignoring the more tangential aspects that would emerge during the usual homeopath's interview. Similarly, *combination homeopathy* is interested only in the patient's main symptom, but it relies on mixtures of different remedies that all share the ability to treat this one outstanding symptom. In other words, a patient with migraine would receive a homeopathic mixture of all the remedies that include headache as one of the symptoms that they cure. Another way to prescribe is according to the *doctrine of signatures*, which places less emphasis on the symptoms in the *Materia Medica* and instead looks for a clue, or signature, that indicates that a particular remedy is the one that should be adopted. Therefore a walnut-based remedy would be appropriate for various mind-related disorders, such as stress, because the walnut resembles a brain.

Figure 3
This is the entry for the homeopathic remedy Aceticum acidum, taken from William Boericke's Materia Medica. It offers numerous curious statements about Aceticum acidum, such as 'Counteracts sausage poisoning' and recommending its use for people worried about business affairs.

Aceticum Acidum
Glacial Acetic Acid
(*Acetic Acid*)

This drug produces a condition of profound anæmia, with some dropsical symptoms, great debility, frequent fainting, dyspnœa, weak heart, vomiting, profuse urination and sweat. Hæmorrhage from any part. Especially indicated in pale, lean persons, with lax, flabby muscles. *Wasting and debility*. Acetic acid has the power to *liquefy albuminous and fibrinous deposits*. Epithelial cancer, internally and locally (W Owens). Sycosis with nodules and formations in the joints. Hard chancre. The 1X solution will soften and cause formation of pus.

Mind	Irritable, worried about business affairs.
Head	Nervous headache, from abuse of narcotics. Blood rushes to head with delirium. Temporal vessels distended. Pain across root of tongue.
Face	*Pale, waxen, emaciated*. Eyes sunken, surrounded by dark rings. Bright red. Sweaty. Epithelioma of lip. Cheeks hot and flushed. Aching in left jaw-point.
Stomach	*Salivation. Fermentation* in stomach. Intense burning thirst. Cold drinks distress. Vomits after every kind of food. Epi-gastric tenderness. Burning pain as of an ulcer. Cancer of stomach. Sour belching and vomiting. Burning waterbrash and profuse salivation. Hyperchlorhydria and gastralgia. *Violent burning pain in stomach and chest, followed by coldness of skin and cold sweat on forehead*. Stomach feels as if she had taken a lot of vinegar.

Abdomen	Feels as if abdomen was sinking in. Frequent watery stools, worse in morning. *Tympanitic*. Ascites. Hæmorrhage from bowels.
Urine	Large quantities of pale urine. Diabetes, with great thirst and debility (*Phos ac*).
Female	Excessive catamenia. *Haemorrhages after labor*. Nausea of pregnancy. Breasts painfully enlarged, distended with milk. Milk impoverished, bluish, transparent, sour. Anæmia of nursing mothers.
Respiratory	Hoarse, hissing respiration; *difficult breathing; cough when inhaling*. Membranous croup. Irritation of trachea and bronchial tubes. False membrane in throat. Profuse bronchorrhœa. Putrid sore throat (gargle).
Back	Pain in back, *relieved only by lying on abdomen*.
Extremities	Emaciation. Œdema of feet and legs.
Skin	Pale, waxen, œdematous. Burning, dry, hot skin, or bathed in profuse sweat. Diminished sensibility of the surface of body. Useful after stings, bites, etc. Varicose swellings. Scurvy; *anasarca*. Bruises; sprains.
Fever	*Hectic, with drenching night-sweats. Red spot on left cheek. No thirst in fever*. Ebullitions. *Sweat profuse, cold*.
Relationship	Acetic acid is antidotal to all anæsthetic vapors. Counteracts sausage poisoning.
Compare	*Ammon acet* (Profuse saccharine urine, patient is bathed in sweat). *Benzoin oderiferum—Spice-wood (night sweats). Ars*; *China*; *Digitalis*; *Liatris* (*General anasarca in heart* and kidney disease, *dropsy*, and chronic diarrhœa).
Dose	Third to thirtieth potency. Not to be repeated too often, except in croup.

With so many approaches and so many possible remedies, some homeopaths employ specific and peculiar techniques for checking that they have found the appropriate treatment. This can include dowsing, whereby a pendulum is held above a shortlist of possible remedies. The direction of swinging should indicate the correct remedy, yet a scientific trial conducted in 2002 showed no evidence for the power of homeopathic dowsing. Six homeopaths were given twenty-six pairs of bottles; one bottle in each pair contained Bryonia remedy and the other contained a placebo, and the challenge was to use dowsing to identify the genuine remedy. Although the homeopaths generally felt that they were selecting with a high degree of confidence, they were successful only 75 times out of 156 trials, which is a success rate of just under 50 per cent: roughly what one would expect from guesswork.

All this ritual – from extreme dilutions to vigorous shaking, from prolonged provings to dubious dowsing – is performed with the ultimate goal of trying to restore a patient's *vital force* to its usual, healthy balance. Hahnemann proposed that this vital force, something akin to the spirit, permeated the body and entirely determined a person's wellbeing. Many modern homeopaths still believe in the crucial significance of the vital force, which in turn means that they tend to reject many of the principles of conventional medicine, such as the role of bacteria as agents of disease. For example, a homeopath would treat a patient with an ear problem by noting every single mental and physical symptom and then prescribing the most appropriate remedy according to the *Materia Medica*; the goal would be to rebalance the patient's vital force. By contrast, a conventional doctor would focus on the patient's main symptoms, perhaps diagnose a bacterial ear infection and then prescribe antibiotics to kill the bacteria.

Not surprisingly, modern science struggles to accept homeopathy. After all, there is no logical reason why like should be guaranteed to cure like; there is no known mechanism that would allow such ultra-weak dilutions (devoid of any ingredient) to impact on our body; and there is no evidence whatsoever to support the existence of a vital force. However, the sheer oddity of homeopathy's philosophy and practice does not necessarily mean that this approach to medicine should be rejected, because the critical test is not how bizarre it is, but whether or not it is effective. This can best be decided via the ordeal of the clinical trial, that tried and trusted tool of evidence-based medicine, which is capable of separating genuine medicine from quackery.

The rise and fall and rise of homeopathy

Homeopathy spread rapidly through Europe during the first half of the nineteenth century, so much so that Hahnemann's philosophy became well established during his own lifetime. The idea that 'like cures like' and the belief that diseases were 'derangements of the spirit-like power that animates the human body' sounded similar to some elements of the still highly respected Greek philosophy of medicine, so homeopathy was greeted with enthusiasm. Moreover, Hahnemann's ideas were emerging before scientists had firmly established the germ theory of disease or the atomic theory of matter, so the vital force and ultra-weak dilutions did not sound quite so strange as they do today.

Signs of Hahnemann's growing influence ranged from opening the world's first homeopathic hospital in Leipzig in 1833 to the use of homeopathy to treat Napoleon's pubic lice. Homeopathy became particularly fashionable in Paris in the 1830s, because Hahnemann set up home in the city after marrying a beautiful Parisian socialite named Marie Mélanie d'Herville-Gohier – he was eighty years old and she was in her early thirties. With her patronage and his reputation, they were able to run a lucrative practice for the wealthy elite, with Mrs Hahnemann assisting her husband in the afternoon and running her own clinic for the poor in the morning.

Elsewhere in Europe, Hahnemann's disciples spread the gospel of homeopathy with their master's voice ringing in their ears: 'He who does not walk exactly on the same line with me, who diverges, if it be but the breadth of a straw to the left or right, is a traitor and I will have nothing to do with him.' Certainly Dr Frederick Quin, who had studied with Hahnemann in Paris, was no such traitor, for he established homeopathy in London in 1827 strictly according to Hahnemann's principles. It soon became highly popular among the British aristocracy, and within half a century it was being practised across the country, with large homeopathic hospitals being founded in London, Bristol, Birmingham, Liverpool and Glasgow.

Although welcomed by many doctors and patients, this rapid growth was not without controversy. When William Henderson, Professor of General Pathology at Edinburgh University, began to support homeopathy in the 1840s, a colleague wrote: 'The consternation manifested by the Medical Faculty in the University and by the College of Physicians was such as might be exhibited in ecclesiastical circles if the Professor of Divinity were to announce that he had become a Mohammedan.'

At roughly the same time, homeopathy was also establishing itself on the other side of the Atlantic. Dr Hans Burch Gram, a Bostonian of Danish descent, learned about homeopathy during a visit to Copenhagen and then brought the idea back to America in 1825. Just as had happened in Britain, homeopathy gained both ardent supporters and fervent critics. The result was that there were 2,500 practitioners and six homeopathic colleges by the outbreak of the American Civil

War, but homeopaths were still largely denied the opportunity to serve in the army. A professor at the Homeopathic Medical College of Missouri argued that this infringed a soldier's right to receive the medical care of his own choice:

Are personal rights abrogated by the Constitution in time of war? Has a soldier no right to think for himself, and to ask for that relief from suffering and death which his experience for years has taught him is best? Has Congress a right to establish a privileged order in medicine in violation of the spirit and genius of our government?

In order to deal with their critics, homeopaths would often point to the successes they had achieved in dealing with major epidemics. As early as 1800, Hahnemann himself had used ultra-dilute Belladonna to combat scarlet fever; then in 1813 he used homeopathy to treat an epidemic of typhus spread by Napoleon's soldiers after their invasion of Russia; and in 1831 homeopathic remedies such as Camphor, Cuprum and Veratrum were apparently successful in central Europe in tackling outbreaks of cholera, a disease that conventional medicine was unable to treat.

This success was repeated during a cholera epidemic in London in 1854, when patients at the London Homoeopathic Hospital had a survival rate of 84 per cent, compared to just 47 per cent for patients receiving more conventional treatment at the nearby Middlesex Hospital. Many homeopaths therefore argued that this was strong evidence in support of homeopathy, because it was possible to construe the results from these two hospitals as the outcome of an informal trial. The percentages allow us to compare the success rates of two treatments on two groups of patients with the same illness, and homeopathic remedies clearly did better than conventional medicine.

However, critics later pointed out three major reasons why

these percentages did not necessarily mean that homeopathy was effective. First, the patients at the two hospitals had the same illness, but that does not necessarily mean that the two hospitals were competing on a level playing field. It could be, for instance, that the patients who attended the London Homoeopathic Hospital were wealthier, which would mean that they were in a better state of health before catching cholera and were better fed and cared for after leaving hospital – all of this, rather than the homeopathic treatment itself, might account for the higher success rate.

Second, as well as differing in the treatment that they offered, the two hospitals may have differed in other important ways. For instance, the London Homoeopathic Hospital might have had a higher standard of hygiene than the Middlesex Hospital, which could easily explain its superior survival rate. After all, we are dealing with an infectious disease, so clean wards, uncontaminated food and safe water were of the utmost importance.

Third, perhaps the higher survival rate at the London Homoeopathic Hospital was not indicative of the success of homeopathy, but rather it pointed to the failure of conventional medicine. Indeed, medical historians suspect that patients who received no medical care would probably have fared better than those who received the conventional medications given at the time. This might seem surprising, but the 1850s still belonged to the era of so-called 'heroic medicine', when doctors probably did more harm than good.

'Heroic medicine' was a term invented in the twentieth century to describe the aggressive practices that dominated healthcare up until the mid-nineteenth century. Patients had to endure bloodletting, intestinal purging, vomiting, sweating and blistering, which generally stressed an already weakened body. On top of this, patients would receive large doses of medications, such as mercury and arsenic, which scientists now know to be highly toxic. The extreme bloodletting suffered by George Washington, as described in Chapter 1, is a prime example of heroic medicine and its harmful impact on a patient. The label 'heroic medicine' reflected the role played by the supposedly heroic doctor, but anyone who survived the treatment was the real hero.

The richest patients were the most heroic of all, because they endured the most severe treatments. This observation was made as early as 1622, when a Florentine physician, Antonio Durazzini, reported on the recovery rates from a fever that was spreading through the region: 'More of those who are able to seek medical advice and treatment die than of the poor.' It was during this period that Latanzio Magiotti, the Grand Duke of Florence's own doctor, said, 'Most Serene Highness, I take the money not for my services as a doctor but as a guard, to prevent some young man who believes everything he reads in books from coming along and stuffing something down the patients which kills them.'

Although the desperate, wealthy and sick continued to rely on doctors, many onlookers openly criticized their practices. Benjamin Franklin commented, 'All drug doctors are quacks,' while the philosopher Voltaire wrote, 'Doctors are men who prescribe medicines of which they know little, to cure diseases of which they know less, in human beings of whom they know nothing.' He advised that a good physician was one who amused his patients while nature cured the disease. These concerns about medicine were also reflected by several dramatists, including Shakespeare, who in *Timon of Athens* has Timon advise: 'Trust not the physician; His antidotes are poisons.' Similarly, in *Le Malade imaginaire*, Molière wrote: 'Nearly all men die of their remedies and not of their illnesses.'

Hence, if no treatment at all would have been better than conventional heroic medicine for cholera patients, then modern sceptics are not surprised that homeopathy was also better than conventional heroic medicine. After all, the sceptics feel that the homeopathic remedies were so diluted that taking them was the equivalent to having no treatment.

In short, we can conclude two things about a patient seeking treatment before the twentieth century. First, the patient would have been better off opting for no treatment rather than heroic medicine. Second, the patient would have been better off opting for homeopathy rather than for heroic medicine. The important question, however, was whether homeopathy was any better than a lack of treatment? Those who supported homeopathy were convinced by their own experience that it was genuinely effective, whereas sceptics argued that such dilute remedies could not possibly benefit the patient.

The arguments continued throughout the nineteenth century; and despite the initially positive response from the aristocracy and significant sections of the medical community, there was a gradual swing against Hahnemann's ideas as each decade passed. For example, the American physician and writer Oliver Wendell Holmes accepted that conventional medicine had failed in the past ('If all the medicine in the world were thrown into the sea, it would be bad for the fish and good for humanity'), but he was not prepared to tolerate homeopathy as the way forward. He called homeopathy 'a mangled mass of perverse ingenuity, of tinsel erudition, of imbecile incredibility and of artful misrepresentation.'

In 1842, Holmes delivered a lecture entitled 'Homeopathy and Its Kindred Delusions', in which he reiterated why Hahnemann's ideas did not make sense from a scientific point of view. He focused particularly on the extreme dilutions at the heart of homeopathy. One way to think about these dilutions is to consider the key ingredient being dissolved in ever greater volumes of liquid. Each time homeopaths dilute the active ingredient by a factor of 100, they are effectively dissolving it in a volume of water or alcohol that is 100 times bigger, and they do this over and over again. Holmes used a calculation by the Italian physician Dr Panvini to explain the bizarre consequences of such repeated dilutions when applied to a starting ingredient of one drop of Chamomile:

Oliver Wendell Holmes

For the first dilution it would take 100 drops of alcohol. For the second dilution it would take 10,000 drops, or about a pint. For the third dilution it would take 100 pints. For the fourth dilution it would take 10,000 pints, or more than 1,000 gallons, and so on to the ninth dilution, which would take ten billion gallons, which he computed would fill the basin of Lake Agnano, a body of water two miles in circumference. The twelfth dilution would of course fill a million such lakes. By the time the seventeenth degree of dilution should be reached, the alcohol required would equal in quantity the waters of ten thousand Adriatic seas. Swallowers of globules, one of your little pellets, moistened in the mingled waves of one million lakes of alcohol, each two miles in circumference, with which had been blended that one drop of Tincture of Camomile, would be of precisely the strength recommended for that medicine in your favorite Jahr's Manual, against the most sudden, frightful, and fatal diseases!

In the same spirit, William Croswell Doane (1832–1913) also took a swipe at homeopathy. As the first Episcopal Bishop of Albany, New York, he penned a piece of doggerel entitled 'Lines on Homoeopathy':

> Stir the mixture well
> Lest it prove inferior,
> Then put half a drop
> Into Lake Superior.
> Every other day
> Take a drop in water,
> You'll be better soon
> Or at least you oughter.

In Europe Sir John Forbes, Queen Victoria's physician, called homeopathy 'an outrage to human reason', a view that was consistent with the entry for homeopathy in the 1891 edition of the *Encyclopaedia Britannica*: 'Hahnemann's errors were great . . . He led his followers far out of the track of sound views of disease.'

Part of the reason for homeopathy's decline in popularity was that the medical establishment was transforming itself from heroic and dangerous into scientific and effective. Clinical trials, such as those that exposed the dangers of bloodletting, were steadily differentiating between hazardous procedures and effective cures. And, as each decade passed, there was an increased understanding of the true causes of disease. One of the most important medical breakthroughs took place during the previously mentioned 1854 London cholera epidemic.

The disease had first hit Britain in 1831, when 23,000 people died; this was followed by the 1849 epidemic, which killed 53,000. During the 1849 epidemic the obstetrician Dr John Snow questioned the established theory that cholera was spread through the air by unknown poisonous vapours. He had been a pioneer of anaesthesia and had administered chloroform to Queen Victoria during the birth of Prince Leopold, so he knew exactly how gaseous poisons affected groups of people; if cholera was caused by a gas, then entire populations should be affected, but instead the disease seemed to be selective about its victims. Therefore, he posited the radical theory that cholera was caused by contact with contaminated water and sewage. He put his theory to the test during the next cholera outbreak in 1854. In London's Soho, he made an observation that seemed to support his theory:

> Within 250 yards of the spot where Cambridge Street joins Broad Street there were upwards of 500 fatal attacks of cholera in 10 days. As soon as I became acquainted with the situation and extent of this eruption of cholera, I suspected some contamination of the water of the much-frequented street-pump in Broad Street.

To investigate his theory he plotted the location of every death on a map of Soho (see Figure 4) and, sure enough, the suspicious pump was at the epicentre. His theory was further backed by his observation that a local coffee shop that served water from the pump had nine customers who had contracted cholera. On the other hand, a nearby workhouse with its own well had no cases, and employees at the brewery on Broad Street had been unaffected because they drank their own produce.

A key piece of evidence was the case of a woman who died of cholera, even though she lived far from Soho. Snow learned, however, that she had previously lived in Soho and had such a fondness for the sweet pump water that she had specially asked for some Broad Street water to be brought to her house. Based on all these observations, Snow persuaded town officials to take the handle off the pump, which halted the supply of contaminated water and brought an end to the cholera

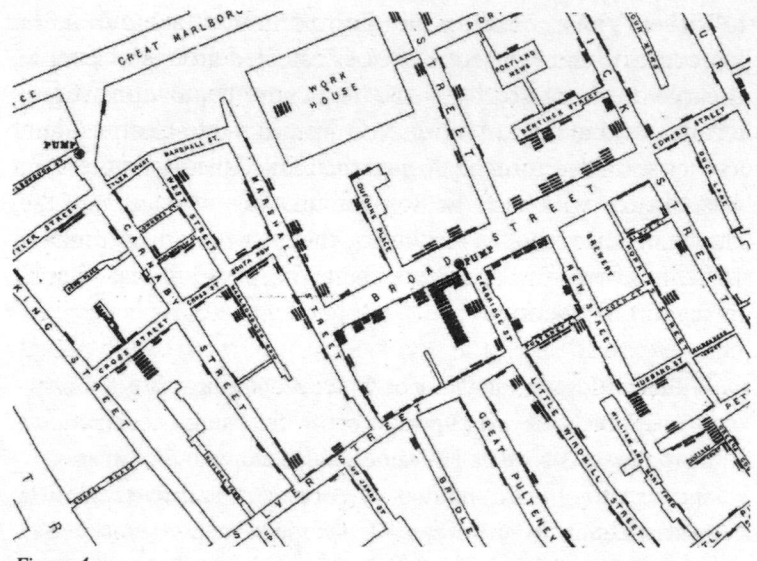

Figure 4
John Snow's map of cholera deaths in Soho, 1854. Each black oblong represents one death, and the Broad Street pump can be seen at the centre of the epidemic.

outbreak. Snow, arguably the world's first epidemiologist, had demonstrated the power of the new scientific approach to medicine, and in 1866 Britain suffered its last cholera outbreak.

Other major scientific breakthroughs included vaccination, which had been growing in popularity since the start of the 1800s, and Joseph Lister's pioneering use of antiseptics in 1865. Thereafter Louis Pasteur invented vaccines for rabies and anthrax, thus contributing to the development of the germ theory of disease. Even more importantly, Robert Koch and his pupils identified the bacteria responsible for cholera, tuberculosis, diphtheria, typhoid, pneumonia, gonorrhoea, leprosy, bubonic plague, tetanus and syphilis. Koch deservedly received the 1905 Nobel Prize for Medicine for these discoveries.

Without any comparable achievements attributed to homeopathy, and without any rigorous evidence or scientific rationale to support it, the use of these ultra-dilute homeopathic remedies continued to decline into the twentieth century in both Europe and America. For example, American homeopathy was dealt a severe blow in 1910 when the Carnegie Foundation asked Abraham Flexner to investigate ways of establishing higher standards for the admission, teaching and graduation of medical students. One of the key recommendations of the Flexner Report was that medical schools should offer a curriculum based on mainstream conventional practice, which effectively ended the teaching of homeopathy in major hospitals.

Homeopathy continued its steady decline, and by the 1920s it seemed that it was destined to become extinct around the world. Then, in 1925, there was a sudden and unexpected revival in Germany, the country where homeopathy had been invented. The man behind the resurgence was an eminent surgeon called August Bier, who used the homeopathic principle of 'like cures like' to treat bronchitis with ether and to cure boils with sulphur. His patients responded well, so he wrote up his findings in a German medical journal. This was the only paper on the subject of homeopathy to be published in Germany in 1925, but it triggered forty-five papers discussing homeopathy the following year, and over the next decade there was a renewed enthusiasm for the potential of ultra-dilute medicines.

This was a timely development for the Third Reich, whose leaders sought to develop the *Neue Deutsche Heilkunde* (the New German Medicine), an innovative medical system that would combine the best of both modern and traditional medicine. The first hospital to implement fully the *Neue Deutsche Heilkunde* was founded in Dresden in 1934 and was named after Rudolf Hess, who was Hitler's deputy at the time.

Hess was strongly in favour of incorporating homeopathy within the *Neue Deutsche Heilkunde*, partly because he believed it to be highly effective, and partly because it had been invented by a German. Furthermore, he viewed homeopathic remedies, most of which were cheap to manufacture, as a low-cost solution to meeting the needs of German healthcare.

Meanwhile, the German Ministry of Health was keen to test whether or not homeopathy was genuinely effective. The Third Reich's chief medical officer, Dr Gerhard Wagner, instigated an unprecedented programme of research, which involved sixty universities and cost hundreds of millions of Reichsmarks. The research effort started immediately after the 1937 Homeopathic World Congress in Berlin and it continued for the next two years, with a particular focus on treating tuberculosis, anaemia and gonorrhoea. The team behind the homeopathy research project included pharmacologists, toxicologists and, of course, homeopaths, who together designed a series of detailed trials and then implemented them rigorously. It is worth noting that those involved in the trials were among the most respected people in their fields, and they maintained the highest ethical and scientific standards in their research.

The results were about to be announced in 1939, but the outbreak of the Second World War prevented publication. The original documents survived the war and were discussed again when the senior researchers reconvened in 1947, but unfortunately their conclusions were never formally announced. Worse still, the documents have never been seen again. It seems that the results of the first comprehensive study of homeopathy have been concealed, lost or destroyed.

Nevertheless, there exists one very detailed account of the Nazi research programme, which was written by Dr Fritz Donner and published posthumously in 1995. Donner had joined the Stuttgart Homeopathic Hospital in the mid-1930s and had contributed to the national research programme in his capacity as a practising homeopath. According to Donner, who claims to have seen all the relevant documents, none of the trials gave any indication in favour of the efficacy of homeopathy: 'It is unfortunately still not generally known that these comparative studies in the area of infectious diseases such as scarlet fever, measles, whooping cough, typhus etc generated results which were not better for homeopathy than for placebo.' He also added: 'Nothing positive emerged from these tests ... except the fact that it was indisputably established that the views [of homeopaths] were based on wishful thinking.'

If Donner was correct, then his statement would be a damning indictment of homeopathy. The first comprehensive and rigorous programme to test the claims of homeopathy, conducted by researchers who were sympathetic to the philosophy and who were to some extent under pressure to prove its validity, had arrived at a wholly negative conclusion. Of course, we cannot be sure that Donner's report was accurate, as the vital documents have never resurfaced. It would, therefore, be wrong to condemn homeopathy based on the testimony of one man's view of research conducted seventy years ago. But even if we ignore the supposed negative results of the Nazi research programme, it is still interesting to note that between Hahnemann's initial research and the end of the Second World War, a period of some one and a half centuries, nobody succeeded in publishing any conclusive scientific evidence to support the notion of homeopathy.

Nature's miracle

After the Second World War, mainstream medicine in America and Europe continued its relentless progress, thanks to further important scientific breakthroughs such as antibiotics. Meanwhile the homeopathic tradition was managing to survive only with the patronage of some powerful and sympathetic supporters. For example, George VI was a fervent believer, so much so that he even named one of his horses Hypericum, after the homeopathic remedy based on St John's wort; the horse went on to win the One Thousand Guineas at Newmarket in 1946. Two years later, King George played an influential role in enabling homeopathic hospitals to come under the umbrella of the newly formed National Health Service.

In America, it was the influence of men like Senator Royal Copeland that allowed homeopathy to survive despite the general trend away from Hahnemann's philosophy and towards the use of treatments with a more scientific and reliable foundation. Both a homeopath and a politician, Copeland successfully persuaded his colleagues that the 1938 Federal Food, Drug, and Cosmetic Act should include the Homeopathic Pharmacopeia of the United States. The Act was supposed to protect patients from unproven or disproven remedies, and yet the claims of homeopathy were still based merely on anecdote and Hahnemann's preaching. So, by including the entire homeopathic catalogue, the Act was giving undue credence to remedies that had no scientific basis.

In India, homeopathy was not only surviving, but it was actually thriving at every level of society, and this success had nothing to do with political manoeuvring or royal patronage. Homeopathy had been introduced there in 1829 by Dr Martin Honigberger, a Transylvanian physician who joined the court of Maharajah Ranjit Singh in Lahore. The idea then spread rapidly throughout India, prospering largely because it was perceived as being in opposition to the imperialist medicine practised by the British invaders. Attitudes towards British medicine were so negative, in fact, that vaccination programmes and attempts to quarantine plague casualties both failed dismally in the mid-nineteenth century.

Moreover, Indians who wanted to pursue a career in conventional medicine often encountered prejudice when they attempted to join the Indian Medical Service, so a more realistic (and cheaper) career option was to train to be a homeopathic practitioner. It was also felt that homeopathy and the Hindu Ayurvedic system of medicine could work together in harmony, and there were even rumours that Hahnemann himself had studied traditional Indian medicine.

As the decades passed, tens of millions of Indians grew to rely solely on homeopathy for their healthcare. And, having imported homeopathy from the West, India then exported it back to the West in the 1970s. At a time when Western patients were looking to the East for alternative systems of medicine – such as acupuncture and Ayurvedic therapies – they also began to embrace homeopathy once again. It was considered by many Westerners to be an exotic, natural, holistic and individualized form of medicine, and an antidote to the corporate medicine being peddled by giant pharmaceutical corporations in Europe and America.

Meanwhile, Western scientists continued to scoff. There were a few scientific trials examining the benefits of homeopathy in the 1950s, 1960s and 1970s, but they were so flimsy that the results were unreliable. In short, there was still no sound evidence to support the idea that such ultra-dilute solutions could act as meaningful medicines. Therefore scientists still considered it absurd that any medical system could be built upon this principle.

Scientists even began to poke fun at homeopaths. For example, because homeopathic liquid remedies are so dilute that they often contain only water, scientists would sarcastically endorse their use for the treatment of one particular medical condition, namely dehydration. Or they would jokingly offer to make each other a drink of homeopathic coffee, which was presumably incredibly diluted and yet tasted incredibly strong, because homeopaths believe that lower amounts of active ingredient are associated with greater potency. Similar logic also implied that a patient who forgot to take a homeopathic remedy might die of an overdose.

Homeopaths accepted that repeated dilution inevitably removes the presence of the active ingredient, and sure enough chemical analysis has always confirmed that 'high-potency' homeopathic remedies are based on nothing more than pure water. Homeopaths were adamant, however, that this water was special because it had a memory of the active ingredient that it once contained. This caused the Australian Council Against Health Fraud to make fun of homeopathy by pointing out that this memory must be highly selective: 'Strangely, the water offered as treatment does not remember the bladders it has been stored in, or the chemicals that may have come into contact with its molecules, or the other contents of the sewers it may have been in, or the cosmic radiation which has blasted through it.'

Then, in June 1988, the laughing suddenly stopped. *Nature*, arguably the most respected science journal in the world, published a research paper with the snappy title 'Human basophil degranulation triggered by very dilute antiserum against IgE'. It took a little deciphering before non-specialists could appreciate the significance of the paper, but very rapidly it became clear that here was a piece of research that seemed to back up some of the claims of homeopaths. If the paper was correct, then ultra-dilute solutions that did not contain any active ingredient did indeed have an impact on biological systems. This could only be possible if the ingredient had left a memory of itself in the water. In turn, such a discovery would imply that homeopaths might have been right all along.

This piece of research, which has become the most famous experiment in the history of homeopathy, was conducted by a charismatic French scientist named Jacques Benveniste, a former racing driver who had taken up medical research after suffering a back injury. Although he published several important scientific papers on a variety of subjects during the course of his career, he would ultimately be remembered only for his *Nature* paper on homeopathy, which shocked the scientific establishment and made headlines around the world.

Benveniste's controversial paper had surprisingly humble beginnings. The research began when one of his colleagues was looking at how *basophils*, a type of white blood cell, reacted to a particular allergen. This is akin to the allergic reaction that might be experienced when pollen hits the eye, but on a much smaller scale. Benveniste's chosen allergen was supposed to be only mildly diluted, but the technician accidentally created a solution so dilute that it was devoid of the allergen. Nevertheless, the technician was stunned to find that the solution still had a significant impact on the basophils. Benveniste was equally astonished, so he asked for the unplanned ultra-dilution experiment to be repeated. Again, the basophils seemed to react to an allergen that was no longer in the solution. Benveniste was not aware of homeopathy at the time, but it was not long before someone pointed out that his experiments were demonstrating the sorts of effects that homeopaths had been championing for two centuries. The results implied that water had some kind of memory of what it had previously contained, and that this memory could have a biological impact. It was such a weird conclusion that Benveniste later commented, 'It was like shaking your car keys in the Seine at Paris and then discovering that water taken from the mouth of the river would start your car!'

The French team continued researching the idea of water memory for another two years. Throughout this period they achieved consistently positive results. For the first time ever, homeopaths could argue that here was scientific evidence to support the mechanisms underlying homeopathy.

Previously, supporters of homeopathy had been forced to rely on arguments that were far from convincing. For example, homeopaths would argue that homeopathy worked in a similar way to vaccination. Vaccination is also a treatment whereby tiny amounts of what causes an illness can be used to combat that illness. At first this seems persuasive, but there is a major difference between homeopathy and vaccination. The amounts of active ingredient used in vaccines might be tiny, perhaps just a few micrograms, but this is still vast compared to a homeopathic remedy. A vaccine contains billions of viruses or virus fragments, whereas most homeopathic remedies do not contain a single molecule of the active ingredient. The flawed analogy between vaccines and homeopathy has been promoted by homeopaths since the nineteenth century, when Oliver Wendell Holmes rebutted it by pointing out that it was akin to 'arguing that a pebble may produce a mountain, because an acorn can become a forest'.

Having satisfied himself that his research findings were valid, Benveniste sent a paper describing his experiments to John Maddox, editor of *Nature*. Maddox duly had the paper refereed, which is a standard procedure that allows independent scientists to check any new results and discuss whether or not the research has been conducted properly. The experimental protocol seemed to be in order, but the claims in the paper were so extraordinary that Maddox took the step of adding a disclaimer alongside the published paper. The last time that Maddox had adopted this unusual approach was in 1974, when he published a paper about Uri Geller's supposed spoon-bending powers. The disclaimer for Benveniste's paper read: 'Editorial reservation: Readers of this article may share the incredulity of the many referees . . . *Nature* has therefore arranged for independent investigators to observe repetitions of the experiments.'

In other words, *Nature* decided to publish Benveniste's work, but with the caveat that the journal would re-check the research by sending a team of experts to visit the French laboratory. The team was led by Maddox himself, and he was joined by Walter Stewart (a chemist) and James Randi (a magician). Randi's inclusion raised some eyebrows, but he had an international reputation for debunking extraordinary claims and uncovering scientific fraud. To illustrate his attitude, Randi would often explain that if his neighbour claimed to have a goat in the garden then he would probably believe him, but if the neighbour said he had a unicorn then Randi would probably want to check how firmly its horn was attached. Randi had established himself as one of the world's leading sceptics back in 1964, when he hit the headlines by offering a reward of $10,000 to anyone who could prove the existence of any paranormal phenomenon, which included therapies such as homeopathy that are contrary to the principles of science. The prize fund had steadily increased to $1 million by 1988, so if the team endorsed Benveniste's result then it would lead to Randi writing out a very large cheque to the Frenchman.

The investigation started within a week of the paper's publication. It lasted four days and involved replicating the key experiment, with Maddox, Stewart and Randi monitoring every stage and checking for flaws in the procedure. They observed the handling of several test tubes containing basophil blood cells, some of which were treated with the homeopathic allergen solution, while the rest were treated with plain water and acted as a control. The task of analysing the test tubes was given to Elisabeth Davenas, Benveniste's assistant, and yet again the result was the same as it had been for the last two years. More of the homeopathically treated cells showed an allergic response than the control cells, implying that the homeopathic solution had genuinely triggered a reaction in the blood cells. Even though the homeopathic solution no longer contained any allergen, its 'memory' of the allergen seemed to be having an impact. The experiment had been successfully replicated.

The investigators, however, were still not convinced. When Davenas analysed the test tubes she knew exactly which ones had been treated with the homeopathic solution, so the investigators were concerned that her analysis might have been deliberately or unconsciously biased. In Chapter 2 we discussed the issue of blinding, which means that patients in a trial should not be aware of whether they are receiving the real treatment or the placebo control treatment. Blinding is equally applicable to doctors and scientists. They should not be aware of whether they are administering or studying the real or the control treatment. The aim of blinding is to minimize bias, and to avoid anybody being influenced by their expectations.

Consequently, the *Nature* team requested Davenas to repeat the analysis, but only after they had blinded her to the contents of the test tubes. Maddox, Randi and Stewart went into a separate room, blanked out the windows with newspapers, removed the labels from the test tubes and replaced them with secret codes that they would later use to identify which samples had been treated with homeopathic solution and which had been treated with water. Davenas repeated her analysis, while colleagues from around the laboratory gathered to await the final result. The Amazing Randi, as he is known on stage, amused the crowd with a few card tricks to help ease the tension.

Eventually Davenas completed her analysis. The secret codes were revealed and the *Nature* team identified which test tubes had been treated homeopathically. This time the results showed that the basophils in the homeopathically treated samples had not reacted differently from the control basophils treated with plain water. The experiment had failed to demonstrate the sort of effects that Benveniste had been finding for the last two years. The results showed no evidence to support homeopathy, and instead they were in line with conventional scientific thinking and all the known laws of physics, chemistry and biology. Some of Benveniste's colleagues burst into tears at the announcement.

Subsequently it emerged that Benveniste had never personally conducted any of the experiments, but had always left everything to Davenas. Moreover, she had always conducted the analysis in an unblinded manner. This meant that it was highly likely that she had accidentally and consistently introduced biases into the results, particularly as she herself was already a strong believer in the power of homeopathy and was keen to prove its efficacy.

When *Nature* published the results of its investigation, the journal pointed out several problems with Benveniste's approach to research. These criticisms included statements such as: 'We believe that experimental data have been uncritically assessed and their imperfections inadequately reported.' Moreover, the journal highlighted the fact that two of the researchers who had contributed to Benveniste's original paper had been partially funded by a French homeopathic company with an annual turnover of over €100 million. Corporate funding is not necessarily problematic, but such potential conflicts of interest had not been formally disclosed. Despite these criticisms, the *Nature* investigators were keen to stress that they were not accusing Benveniste of deliberate fraud, but merely that he and his team were deluding themselves and had not been conducting their experiments rigorously.

A lack of rigour, particularly a lack of blinding, can seriously bias any scientific result even for the most honest and well-intentioned scientist. Imagine the following scenario: a scientist has staked his reputation on the hypothesis that men have superior spatial awareness and motor skills, and he thinks he can demonstrate that this is the case by inviting men and women to draw freehand circles and then comparing the quality of their drawings. The experiment begins – the men and women draw their circles, they write their names at the top of the papers, the drawings are collected by an assistant and handed to the scientist, who judges the circles by eye and gives each one marks out of ten. However, because he can see the names of the artists at the top of each drawing, he might be tempted subconsciously to mark the men's circles more generously. Consequently, regardless of the truth, it is more likely that the resulting data would support his hypothesis that men are better than women at drawing circles. By contrast, if the experiment were to be repeated and the artists were given numbers to disguise their gender temporarily, then the prejudiced scientist becomes blinded and is more likely to give a fairer assessment of each circle. The new result is likely to be more reliable.

In the Benveniste case, the problem was that Davenas was unblinded and prejudiced in favour of homeopathy, and this combination of factors could have biased her results. In particular, Davenas's experiments required her to judge whether a homeopathic preparation caused blood cells to exhibit an allergic reaction, which is not a clear-cut decision, even when the cells are viewed through a microscope. Indeed, judging the extent of a cell's allergic reaction is similar to examining a circle's roundness: both are equally prone to personal interpretation and bias.

For example, Davenas would have come across many borderline cases – has the cell undergone an allergic reaction or not? There might have been a subconscious temptation to judge such borderline cells as exhibiting allergic reactions if she knew that they had been treated homeopathically. Or she might have been subconsciously tempted to give the opposite judgement if she knew that they had been treated with plain water. However, by asking Davenas to repeat the experiment without any labels on the test tubes, the *Nature* investigators ensured that she was blind and unbiased in her decisions; whereupon the homeopathic solutions and water led to similar results. A fair test had shown that the homeopathic solutions had no impact on the basophil cells.

While Benveniste readily accepted some elements of the criticism, he steadfastly defended the core of his research and argued that the results that he had accumulated over the course of two years could not be negated by what the *Nature* team had observed in just a few days. He explained that the mistakes that had been witnessed by Maddox, Randi and Stewart were caused by the unusual circumstances, namely that his team was working under intense pressure and in the media spotlight.

Benveniste remained convinced that his work would ultimately be recognized with a Nobel Prize, but instead he was merely rewarded with a satirical award known as the Ig Nobel Prize. In fact, he won an Ig Nobel Prize in 1991 and then another one in 1998, making him the first scientist to win two Ig Nobels. As the years passed, Benveniste saw his scientific reputation decline in the press and among his peers, which led him to complain that he was being victimized. He even compared himself to Galileo, because they had both been subjected to attacks when they dared to speak out against the establishment. This was a flawed comparison for two major reasons. First, Galileo was attacked largely by the religious establishment, rather than by his scientific peers. Second, Galileo was in a different class to Benveniste – after all, Galileo's observations stood up to scrutiny and his experimental results were replicated by others.

Benveniste struggled to retain his academic post as a result of the *Nature* debacle, but he was determined not to abandon his research, so he established a company called DigiBio to

nurture and promote his ideas. Among their wilder pronouncements, researchers at DigiBio stated that not only could water
hold a memory of what it had previously contained, but that
this memory could also be digitized, transmitted via email and
reintroduced into another sample of water, which in turn could
affect basophil cells. Although Benveniste died in 2004,
DigiBio has continued its campaign to have his ideas taken
seriously. Its website proclaims:

> From the first high dilution experiments in 1984 to the present,
> thousands of experiments have been made, enriching and con-
> siderably consolidating our initial knowledge. Up to now, we
> must observe that not a single flaw has been discovered in
> these experiments and that no valid counter-experiments have
> ever been proposed.

In fact, within a year of Benveniste's original 1988 paper,
Nature had published three papers by scientists who failed to
reproduce the supposed effect of ultra-dilute solutions. Even
the U.S. Defense Advanced Research Projects Agency
(DARPA) collaborated with homeopaths to test DigiBio's
claim that Benveniste's effects could be digitized and sent via
email, but they came to the following conclusion: 'Our team
found no replicable effects from digital signals.'

On the other hand, there have been occasional papers that
claim to replicate the sort of effects observed by Benveniste,
but so far none of them has consistently or convincingly presented the sort of evidence that would posthumously vindicate
the Frenchman. In 1999, Dr Andrew Vickers looked at 120
research papers related to Benveniste's work and other types
of basic research into the actions of homeopathic remedies. At
the time, he was based at the Royal London Homoeopathic
Hospital, so he was certainly open-minded about the potential
of homeopathy. Yet Vickers was struck by the failure of

independent scientists to replicate any homeopathic effect: 'In the few instances where a research team has set out to replicate the work of another, either the results were negative or the methodology was questionable.' Independent replication is a vital part of how science progresses. One single set of experiments can be wrong for a range of reasons, such as lack of rigour, fraud or just bad luck, so independent replication is a way of checking (and re-checking) that the original discovery is genuine. Benveniste's research had failed this test.

Indeed, James Randi has continued to offer his $1 million to anyone who can independently reproduce the effects claimed by Benveniste. BBC television took up the challenge as part of its *Horizon* science documentary series, gathering together a team of scientists to oversee the project. They examined the effect of a homeopathically diluted *histamine* on cells, and compared this with the effect of pure water. Histamine is associated with allergic responses in cells, but would it still cause cells to react if it had been diluted to the extent that it was no longer present? Professor Martin Bland of St George's Hospital Medical School announced the final result: 'There's absolutely no evidence at all to say that there is any difference between the solution that started off as pure water and the solution that started off with the histamine.' As anecdotal evidence to reinforce the point, Randi mentioned the following story during the programme: 'I also consumed sixty-four times the prescribed dosage of homeopathic sleeping pills and didn't even feel drowsy. I did this before a meeting of the US Congress – if that doesn't put you to sleep, nothing will.'

While biologists were trying and failing to find evidence for homeopathy acting at a fundamental cellular level, physicists tried to examine homeopathy at a basic molecular level. It was clear that ultra-dilute homeopathic solutions contained only water and no molecules of the active ingredient, but some physicists wondered if the water molecules somehow had altered their arrangement in order to retain a memory of the earlier ingredient.

Over the last two decades, physicists have published the results of dozens of experiments examining the molecular structure of normal water versus homeopathically prepared water. They have used powerful and arcane techniques such as nuclear magnetic resonance (NMR), Raman spectroscopy and light absorption to look for the slightest evidence that water has a memory of what it once contained. Unfortunately, a review of these studies published in *The Journal of Alternative and Complementary Medicine* in 2003 showed that these experiments were generally of poor quality and prone to errors.

For example, one NMR experiment claimed to detect a difference between the molecules in ordinary water and those in a homeopathic remedy, but in the end this was attributed to a problem with the equipment. The NMR apparatus is supplied with test tubes made of soda glass, which is not a very stable form of glass. Hence, when the homeopathic solution was shaken during its preparation, glass molecules were leached into the solution. Not surprisingly, this homeopathic solution responded differently to the pure water in terms of its NMR profile, which initially gave the misleading impression that the homeopathic solution was demonstrating a water-memory effect. Sure enough, when another research team repeated the experiment with borosilicate glass test tubes, which are much more stable than soda glass, the NMR instrument could no longer detect any difference between water and homeopathic remedies. Yet again, experiments have so far failed to find anything surprising about the behaviour of molecules in homeopathic solutions.

In summary, homeopaths have been disappointed that physicists probing water molecules have found nothing special about homeopathic remedies. Similarly, biologists looking at single cells have not made any great breakthroughs in finding convincing evidence that might support homeopathy.

All of this, however, matters very little in terms of the main homeopathy debate, because what happens at the molecular or cellular level is of much less interest than what happens to patients. Forget biology or physics, because homeopathy is all about medicine. The fundamental question is straightforward: does homeopathy heal patients?

Homeopaths, of course, have always been confident that their remedies cured a range of symptoms, but in order to persuade doctors and everyone else that homeopathy was truly effective, they needed concrete evidence from scientific trials. We have explained in previous chapters that the most conclusive type of clinical trial is the randomized, placebo-controlled, double-blind trial; if such trials could generate results that supported Hahnemann's ideas, then this would force the medical establishment to embrace homeopathy. Alternatively, if these studies failed to show that ultra-dilute solutions offered any benefit, then this would mean that homeopathy was nothing more than quackery. As the twenty-first century approached, rigorous trials were about to be conducted on a massive scale. The results would eventually settle the debate over homeopathy once and for all.

Homeopathy on trial

For many homeopaths, the lack of definitive scientific evidence to support their remedies was not a matter of concern, because they could cite numerous examples that seemed to demonstrate the effectiveness of their interventions. For instance, *The Complete Idiot's Guide to Homeopathy* by David W. Sollars includes a story told by a mother who treated her son, Kailin. The boy had burned his arm at a barbecue, but luckily the people hosting the occasion had just purchased a homeopathic home kit:

> I told them to get it quickly as I held my ice-filled glass on his arm, which helped the pain a little. Within a couple of minutes the kit arrived and I chose the remedy Cantharis and gave Kailin a dose. Within two or three minutes the pain stopped and we all watched over the next fifteen minutes as the skin started to lighten in color. I repeated the remedy several times, whenever he said the pain was starting to return. By the next day the burn was all but gone and totally cleared in two days. We were all amazed that no blisters were ever formed.

This case seems impressive, but it was dissected and undermined by Jay W. Shelton, author of *Homeopathy*. He identifies four questions that challenge the significance of this and similar cases. First, this seems like a classic first-degree burn, the least serious type, which causes damage only to the surface of the skin; so should we really be so surprised about the lack of blisters? Second, why should homeopathy get any credit, when the recovery process might have been entirely due to the body's natural healing ability? Third, is it possible that the ice-filled glass played the most important role in helping the child? Finally, if homeopathy did indeed help the patient, then could its influence have been entirely due to the placebo effect? In the previous chapter we saw that the power of placebo is so great that it can make a useless therapy appear like a truly valuable one, as long as a patient has confidence in the therapy.

When scientists dismiss such human cases as partly or wholly due to the placebo effect, homeopaths often cite cases of animal healing, because they believe that animals are immune to the placebo effect. It is true that many pet owners and farmers feel that homeopathy helps their animals, and it is also true that these creatures are oblivious to what a pill is supposed to do, but the value of these anecdotes collapses under closer scrutiny.

For instance, the animal is unaware of which treatment it is receiving or how it is supposed to respond, yet the fact remains that the person monitoring the animal is fully aware. In other words, the animal is effectively blinded to what is happening, but the person reporting the events is unblinded and is therefore unreliable. For instance, an anxious pet owner who has faith in homeopathy might focus on any sign of improvement based on expectation and hope, while ignoring symptoms that have deteriorated. Even if the animal has definitely improved beyond the placebo effect, then this could be due to a variety of factors other than the homeopathic pill, such as the extra care and attention being provided by a caring and concerned owner.

In short, the medical establishment will not accept anecdotal evidence – based on either human or animal patients – as reliable enough to support homeopathy or any other treatment. No amount of anecdote can stand in place of firm evidence, or, as scientists like to say, 'The plural of anecdote is not data.'

Medical scientists place an emphasis on data because the best way to analyse the impact of any therapy is to look at the results from rigorous scientific investigations, particularly clinical trials. By way of a quick recap, you will recall that Chapter 1 revealed the uncanny ability of the randomized clinical trial to show which therapies work and which do not. Chapter 2 then built on that foundation to show how this technique could be used to test the claims of acupuncturists. So what happens when homeopathy is submitted to the same scientific scrutiny?

In theory it should be much easier to test homeopathy than acupuncture, because it is much more obvious how to take into account the placebo effect. A homeopathic trial would require the random assignment of patients into two groups, namely a group treated homeopathically and a placebo control group. The patients would not be told to which group they had been assigned. Both groups would receive an empathic encounter with a homoeopath, who would also be blinded, inasmuch as he or she would not know which patients belonged to which group. Researchers would then create two batches of pills that were identical, except that one batch would have been treated with a drop of homeopathic solution and the other batch would remain plain. The treatment group would receive the homeopathic pill and the control group would receive the plain pill. Patients in both groups should experience some improvement, simply due to the placebo effect. The critical question is this: does the treatment group on average show significant improvements over and above the control group? If the answer is 'yes', then this would clearly indicate that homeopathy is genuinely effective. If, however, the answer is 'no' and each group shows a similar response, then homeopathy would be exposed as having nothing more than a placebo effect.

Before looking at the trials conducted with humans, it is interesting to note that there have been some randomized placebo-controlled trials of homeopathy's impact on animals. The overall conclusion of the major studies is that homeopathy offers no benefit to animals. For example, in 2003 the National Veterinary Institute in Sweden conducted a double-blind trial of the homeopathic remedy *Podophyllum* as a cure for diarrhoea in calves, and it found no evidence for the efficacy of homeopathy. More recently, a Cambridge University research group conducted a double-blind trial to compare homeopathy against a placebo as a treatment for mastitis for 250 cows. An objective way of checking for any improvement in inflammation of the udder is to count the number of white blood cells in the cow's milk, and the conclusion was that homeopathy was no more effective than the placebo.

When scientists looked at the evidence in terms of human

patients, the picture was more complicated. The good news was that by the mid 1990s there had been well over 100 published trials seeking to decide the therapeutic value of homeopathy. The bad news was that this mountain of research consisted largely of poorly conducted trials, often with inadequate randomization, or with no proper control group, or with insufficient numbers of patients. None of these trials was able to give a definitive answer to whether or not homeopathy benefited patients beyond the placebo effect.

With nothing to rely on except unconvincing anecdotes and inconclusive trials, the arguments for and against homeopathy were deadlocked. Then, in 1997, an international research team took a dramatic step towards settling the homeopathy debate. They were led by Klaus Linde, a senior figure at the Munich-based Centre for Complementary Medicine Research. He and his colleagues decided to examine the considerable body of research into homeopathy in order to develop an over-arching conclusion that took into consideration each and every trial. This is known as a *meta-analysis*, which means an analysis of various analyses. In other words, each individual trial into homeopathy concluded with an analysis of its own data, and Linde was proposing to pool all these separate analyses in order to generate a new, more reliable, overall result. Meta-analysis can be considered as a particular type of systematic review, a concept that was introduced in the previous chapter. Like a systematic review, a meta-analysis attempts to draw an overall conclusion from several separate trials, except that a meta-analysis tends to involve a more mathematical approach.

Although the term meta-analysis might be unfamiliar to many readers, it is a concept that crops up in a range of familiar situations where it is important to make sense of lots of data. In the run-up to a general election, for instance, several newspapers might publish opinion polls with conflicting results. In this situation it would be sensible to combine all the data from all the polls, which ought to lead to a more reliable conclusion than any single poll, because the meta-poll (i.e. poll of polls) reflects the complete data from a much larger group of voters.

The power of meta-analysis becomes obvious if we examine some hypothetical sets of data concerning astrology. If your astrological sign determined your character, then an astrologer should be able to identify a person's star sign after an interview. Imagine that a series of five experiments is conducted around the world by rival research groups. In each case, the same astrologer is simply asked to identify correctly a person's star sign based on a five-minute conversation. The experiments range in size from 20 to 290 participants, but the protocol is the same in each case.

Chance alone would give rise to a success rate of one correct identification (or hit) in twelve, so the astrologer would have to do significantly better than this to give credence to the notion of astrology. The five experiments lead to the following success rates:

Experiment 1   12   hits out of   170   (or 0.85 hits out of 12)

Experiment 2   5   hits out of   50   (or 1.20 hits out of 12)

Experiment 3   5   hits out of   20   (or 3.00 hits out of 12)

Experiment 4   6   hits out of   70   (or 1.03 hits out of 12)

Experiment 5   21   hits out of   290   (or 0.87 hits out of 12)

On its own, the third experiment seems to suggest that astrology works, because a hit rate equivalent to 5 out of 20 is much higher than chance would predict. Indeed, the majority of experiments (three out of five) imply a higher than expected hit rate, so one way to interpret these sets of data would be to conclude that, in general, the experiments support astrology. However, a meta-analysis would come to a different conclusion.

The meta-analysis would start by pointing out that the number of attempts made by the astrologer in any one of the experiments was relatively small, and therefore the result of any single experiment could be explained by mere chance. In other words, the result of any one of these experiments is effectively meaningless. Next, the researcher doing the meta-analysis would combine all the data from the individual experiments as though they were part of one giant experiment. This tells us that the astrologer had 49 hits out of 600 in total, which is equivalent to a hit rate of 0.98 out of 12, which is very close to 1 out of 12, the hit rate expected by chance alone. The conclusion of this hypothetical meta-analysis would be that the astrologer has demonstrated no special ability to determine a person's star sign based on their personality. This conclusion is far more reliable than anything that could have been deduced solely from any one of the small-scale experiments. In scientific terms: a meta-analysis is said to minimize random and selection biases.

Turning to medical research, there are numerous treatments that have been tested by meta-analysis. For example, in the 1980s researchers wanted to know if corticosteroid medication could help reduce respiratory problems in premature babies. They designed a trial which involved giving the treatment to pregnant women likely to have premature births and then monitoring the babies born to these mothers. Ideally, the researchers would have conducted one trial in a single hospital with a large number of cases, but it was only possible to identify a few suitable cases each year per hospital, so it would have taken several years to accumulate sufficient data in this manner. Instead, the researchers conducted several trials across several hospitals. The results of each individual trial varied from hospital to hospital, because the numbers of babies in each trial was small and random influences were large. Yet a meta-analysis of all the trials showed with certainty that corticosteroid medication during pregnancy benefited premature babies. This treatment is part of the reason why the number of infant deaths due to respiratory distress syndrome has fallen dramatically – there were 25,000 such deaths in America in the early 1950s and today the number is fewer than 500.

The meta-analysis in the premature baby study was fairly straightforward, because the individual trials were similar to each other and so they could be merged easily. The same is true of the hypothetical example concerning astrology. Unfortunately, conducting a meta-analysis is often a messy business, because the individual trials have generally been conducted in different ways. Trials for the same medication might vary according to the dose given, the period of monitoring, and so on. In Linde's case, the meta-analysis was particularly problematic. In order to draw a conclusion about the efficacy of homeopathy, Linde was attempting to include homeopathy trials investigating a huge variety of remedies, across a range of potencies, being used to treat a wide range of conditions, such as asthma and minor burns.

Linde trawled through the computer databases, attended numerous homeopathic conferences, contacted researchers in the field and eventually found 186 published trials on homeopathy. He and his colleagues then decided to exclude from his meta-analysis those trials that failed to meet certain basic conditions. For example, in addition to a group of patients being treated with homeopathy and a control group of patients, an acceptable trial had to have a placebo for the control-group patients, or there had to be random allocation of patients to the treatment and control groups. This left eighty-nine trials. What followed was months of careful statistical analysis, so that each trial contributed appropriately to the final result. For example, the result of a very small trial would carry very little weight in the overall conclusion, because the reliability of a trial's result is closely linked to the number of participants in the trial.

The meta-analysis was eventually published in September 1997 in the *Lancet*. It was one of the most controversial medical research papers of the year, because its conclusion endorsed exactly what homeopaths had been saying for two centuries. On average, patients receiving homeopathy were much more likely to show signs of improvement than those patients in the control groups receiving placebo. The paper concluded: 'The results of our meta-analysis are not compatible with the hypothesis that the clinical effects of homeopathy are completely due to placebo.' In other words, according to the meta-analysis, homeopathy was genuinely effective.

Not surprisingly, Linde's conclusion was questioned by opponents of homeopathy. Critics argued that his meta-analysis had been too lax, inasmuch as it had included too many trials of relatively poor quality, and they feared that these might have biased the overall conclusion in favour of homeopathy. Homeopaths responded that there had been a quality threshold, which Linde had implemented specifically in order to exclude poor-quality trials. Remember, Linde had included only those trials that were placebo-controlled or randomized. Nevertheless, the critics were still unhappy, as they maintained that the quality threshold had not been set high enough.

Because poorer-quality trials are more likely to produce misleading results, researchers have developed techniques for assessing quality and weeding out those trials that should not be taken seriously. For example, the Oxford quality scoring system, developed in 1996 by Alejandro Jadad and his colleagues at Oxford University, can be used to give a trial a score between 0 points (very poor) and 5 points (rigorous). The system works by awarding or deducting points according to what appears in the published version of the trial. So if the research paper confirms that there was randomization of patients then it receives a point, yet this point can be deducted if the randomization appears to have been inadequate. Or the trial can score a point if the paper describes how the researchers dealt with the data from patients who dropped out from the trial. If the researchers have thought about this in detail and bothered to describe it in their research paper, then it is a good indication of their overall level of rigour.

Critics pointed out that sixty-eight out of the eighty-nine trials in Linde's meta-analysis scored only 3 or less on the Oxford scale, which meant that three-quarters of the trials were substandard. Moreover, critics pointed out that restricting the meta-analysis to the higher-quality trials (4 or 5 points) drastically reduced the apparent efficacy of homeopathy. In fact, the conclusion of the twenty-one higher-quality trials was that homeopathy offered either a small benefit for patients or no benefit at all. Despite the amount of data available from these twenty-one trials, it was still not possible to distinguish between these two possibilities.

In time, Linde and his colleagues agreed that their critics had a valid point, and in 1999 they published a follow-up paper that re-examined the same data with a special emphasis on the quality of the individual trials. Linde wrote: 'We conclude that in the study set investigated, there was clear evidence that studies with better methodological quality tended to yield less positive results.' Then, referring back to the original meta-analysis, he stressed: 'It seems, therefore, likely that our meta-analysis at least over-estimated the effects of homeopathic treatments.'

Linde's original 1997 paper had supported homeopathy, yet

his revised 1999 paper was much more equivocal. His re-analysis of his own meta-analysis obviously disappointed the alternative medicine community, yet it was also frustrating for the medical establishment. Everyone was dissatisfied because Linde was neither able to claim that homeopathy was effective, nor was he able to dismiss it as a mere placebo.

Despite the lack of clear-cut evidence in either direction, the public was increasingly turning towards homeopathy, either consulting practitioners or buying over-the-counter remedies. This gave researchers a renewed sense of urgency to test the therapy via larger, more rigorously conducted trials. Hence, homeopathy was subjected to a much higher level of scrutiny from the late 1990s onwards.

This eventually prompted Dr Aijing Shang and his colleagues at the University of Berne, Switzerland, to under-take a fresh meta-analysis of all the trials published up to January 2003. The medical research group at Berne, which is led by Professor Mathias Egger, has a world-wide reputation for excellence and the Swiss government had provided the team with adequate funding for a fully rigorous meta-analysis. Hopes were high that Shang would at last be able to deliver a reliable conclusion. Indeed, after two centuries of bitter dispute between homeopaths and mainstream medics, Shang's meta-analysis was destined to decide, at last, who was right and who was wrong.

Shang was ruthless in his demand for quality, which meant that his meta-analysis included only those trials with large numbers of participants, decent blinding and proper random-ization. In the end, he was left with only eight homeopathy trials.

After studying the data from these eight trials – the best available trials on homeopathy – his meta-analysis reached its momentous conclusion. On average, homeopathy was only very marginally more effective than placebo. So, did this tiny marginal average benefit suggest that homeopathy actually healed patients?

Before answering this question, it is important to realize that the results of every scientific analysis are always associated with a level of uncertainty. For example, analysing the age of the Earth gives a result of 4,550 million years, and the error is give or take 30 million years. The uncertainty associated with Shang's estimate on the efficacy of homeopathy was such that his conclusion was compatible with the judgement that homeopathy acted as nothing more than a placebo. In fact, the most sensible interpretation of the meta-analysis was that homeopathy was indeed nothing more than a placebo.

This interpretation becomes more convincing if we bear in mind another aspect of his research. While conducting his meta-analysis on homeopathy, he also conducted a meta-analysis for a whole variety of new, conventional pharmaceuticals. These pharmaceuticals had been tested on the same illnesses that had been considered for the homeopathy meta-analysis. In this secondary meta-analysis, Shang scrupulously applied exactly the same selection criteria to these conventional drug trials as he had done in his homeopathy meta-analysis. The result of his meta-analysis on conventional drug trials was that on average they worked. Although this result also had an uncertainty associated with it, the average benefit was so large that the effectiveness of these new conventional drugs was not in any doubt.

The contrast between the homeopathic trials and the conventional drug trials was striking. Homeopathy had failed to show a clear benefit for patients and the result was compatible with homeopathy acting as a placebo, whereas conventional drugs had shown a clear benefit for patients, which suggested that they do indeed have a genuine physiological impact on the body. This illustrates the stark difference between pseudo-medicine and real medicine.

Shang published his results in the *Lancet* in August 2005. Based on his meta-analysis, he concluded: 'This finding is compatible with the notion that the clinical effects of homeopathy are placebo effects.' Reinforcing this point, the *Lancet* ran an editorial entitled 'The end of homeopathy' in which it argued that 'doctors need to be bold and honest with their patients about homeopathy's lack of benefit'. This sparked major news stories around the world, angering homeopaths who refused to accept the conclusions of Shang's meta-analysis and the *Lancet*'s accompanying statement. They attempted to undermine the research by pointing out four key issues, but in fact each of their criticisms can be easily addressed.

1   Homeopaths might argue that Shang's paper indicates a positive effect for homeopathy, and that his meta-analysis therefore supports homeopathy.

There is indeed a positive effect for homeopathy, but it is very small and entirely compatible with the treatment being a placebo. Shang's paper is the most comprehensive analysis of homeopathy during its 200-year history, and such a paucity of positive evidence has to be interpreted as a blow for homeopathy. Crucially, Shang's analysis confirms the results of about a dozen other meta-analyses and systematic reviews published over the last decade, all of which fail to show that homeopathy offers any benefit beyond placebo.

2   Homeopaths claim that Shang dredged the data, which means that the meta-analysis was conducted in such a way as to bias the conclusion.

There are indeed many ways to conduct a meta-analysis. Therefore it is possible to 'dredge the data' in different ways until the most positive or negative result emerges, but importantly Shang had stated what his approach would be before embarking on the meta-analysis, and his approach seemed reasonable and unbiased. In other words, the research was impartial because the goalposts were decided before the data was examined, and the goalposts were of a fair size and were not moved once the research was under way.

3   Homeopaths point out that the meta-analysis included trials for several illnesses, which makes it too crude to say anything meaningful about homeopathy's ability to treat individual conditions.

This over-arching meta-analysis was prompted by the fact that there has been no convincing evidence that homeopathy can treat any individual condition. Whenever researchers have conducted systematic reviews of homeopathy for a particular condition, the results have been consistently disappointing. For headaches and migraine: 'The trial data available to date do not suggest that homeopathy is effective.' For muscle soreness, the most tested condition: 'The published evidence to date does not support the hypothesis that homoeopathic remedies used in these studies are more efficacious than placebo.' For Arnica in the treatment of conditions associated with tissue trauma (e.g. post-operative or post-dental), which is the most widely used homeopathic remedy: 'The claim that homeo-pathic Arnica is efficacious beyond placebo is not supported by rigorous clinical trials.'

4   Homeopaths point out that they offer a highly in-dividualized treatment, which is not suited to large-scale trials in which the homeopathic remedy is standardized.

Indeed, most trials have not been individualized, but there have been trials in which patients were given detailed consultations and either individualized homeopathic prescriptions or placebo. For example, an individualized trial monitoring ninety-eight patients with chronic headaches over the course of twelve weeks led to the conclusion: 'There was no significant difference in any parameter between homeopathy and placebo.' Another trial focused on ninety-six children with asthma and looked at their progress after twelve months of receiving individualized homeopathy or a placebo as an adjunct to their conventional treatment. It concluded: 'This study provides no evidence that adjunctive homeopathic remedies, as prescribed by experienced homeopathic practitioners, are superior to placebo.'

Shang's view of homeopathy is backed up by the Cochrane Collaboration, the highly respected, independent evaluator of medicines introduced in the previous chapter. There are Cochrane reviews on homeopathy for the induction of labour and the treatment of dementia, chronic asthma and flu. Cochrane's conclusions are based on sixteen trials involving over 5,000 patients. Over and over again, the evidence is either non-existent or shaky, leading to conclusions such as 'there is not enough evidence to reliably assess the possible role of homeopathy in asthma'; 'current evidence does not support a preventative effect'; and 'there is insufficient evidence to recommend the use of homeopathy as a method of induction'.

It is interesting to contrast the tenor of these comments on homeopathy with Cochrane's conclusion on a conventional medicine such as aspirin: 'Aspirin is an effective analgesic for acute pain of moderate to severe intensity with a clear dose-response.' Moreover, Cochrane confirms how the efficacy of real medicine is so robust that it can be tested in different ways: 'Type of pain model, pain measurement, sample size, quality of study design, and study duration had no significant impact on the results.' This is the sort of confident conclusion that emerges when a genuinely effective medicine is tested. Sadly, research into homeopathy has failed to deliver any kind of positive conclusion.

Conclusions

It has taken several thousand words to review the history of homeopathy and to survey the various attempts to test its efficacy, but the conclusion is simple: hundreds of trials have failed to deliver significant or convincing evidence to support the use of homeopathy for the treatment of any particular ailment. On the contrary, it would be fair to say that there is a mountain of evidence to suggest that homeopathic remedies simply do not work. This should not be such a surprising conclusion when we recall that they typically do not contain a single molecule of any active ingredient.

This raises an interesting question: with no evidence that it works and with no reason why it ought to work, why is it that homeopathy has grown so rapidly over the last decade into a multi-billion-dollar global industry? Why do so many people think that homeopathy works, when the evidence, frankly, shows that it does not?

One problem is that the public are unaware of the vast body of research that undermines homeopathy. While Linde's original overly optimistic paper from 1997 is hyped on many pro-homeopathy websites, there are far fewer mentions of his more equivocal 1999 re-analysis of exactly the same data. Similarly, the even more important and more negative 2005 paper by Shang is often omitted from homeopathy websites.

Worse still, the public can be misled by news stories that show homeopathy in an unjustifiably sympathetic light. One of the most high-profile homeopathy news stories in recent years concerned a study by the Bristol Homoeopathic Hospital published in 2005. The hospital tracked 6,500 patients during a six-year study and observed that 70 per cent of those suffering with chronic diseases reported positive health changes after homeopathic treatment. As far as the public was concerned, this appeared to be an extraordinarily positive result. However, the study had no control group, so it was impossible to determine whether these patients would have improved without any homeopathic treatment. The 70 per cent improvement rate could have been due to any number of factors, including natural healing processes, or patients being reluctant to disappoint whoever was interviewing them, or the placebo effect, or any other treatments that these patients may have been using. Science writer Timandra Harkness was one of many critics who tried to point out why the Bristol study was largely meaningless: 'It's as if you had a theory that feeding children nothing but cheese made them grow taller, so you fed all your children cheese, measured them after a year and said *There – all of them have grown taller – proof that cheese works!*'

We suggest that you ignore the occasional media hype and instead rely on our conclusion, because it is based on examining all the reliable evidence – and the evidence suggests that homeopathy acts as nothing more than a placebo. For this reason, we strongly advise you to avoid homeopathic remedies if you are looking for a medicine that is more than just make-believe.

Before ending this chapter, it is important to reiterate that we have come to our conclusions about homeopathy based on a fair, thorough, scientific assessment of the evidence. We have no axe to grind and have remained steadfastly open-minded in our examination of homeopathy. Moreover, one of us has had a considerable amount of experience in homeopathy and has even spent time practising as a homeopath. After graduating from a conventional medical school, Professor

Ernst then trained as a homeopath. He even practised at the homeopathic hospital in Munich, treating inpatients for a whole range of conditions. He recalls that the patients seemed to benefit, but at the time it was hard to determine whether this was due to homeopathy, the placebo effect, the dietary advice given by doctors, the body's natural healing ability, or something else.

Ernst continued to practise (and indeed receive) homeopathy for many years, remaining open to its potential. If homeopathy could be shown to be effective, then he and his colleagues would have been overjoyed, as it would offer fresh hope for patients and present new avenues of research in medicine, biology, chemistry and even physics. Unfortunately, as Ernst took a step back and began to look at the research into this form of medicine, he became increasingly disillusioned.

One key piece of research that helped to change Ernst's view was conducted in 1991 by the German pharmacologist Professor W. H. Hopff, who repeated Hahnemann's original experiment with Cinchona – according to Hahnemann, if a medicine that cured malaria was given to a healthy volunteer, then it would actually generate the symptoms of malaria. Using his own students as guinea pigs, the professor compared Cinchona with a placebo and discovered no difference. Neither positive nor negative. In short, Hahnemann's results, which provided the foundation for homeopathy, were simply wrong. Such trials made it clear to Ernst that homeopathic medicines are nothing more than elaborate placebos.

Nevertheless, some readers might still feel that elaborate placebos are perfectly acceptable. You might feel that placebos help patients, and that this alone justifies the use of homeopathy. Some mainstream doctors sympathize with this view, while many others strongly disagree and feel that there are reasons why the placebo effect alone is not enough to justify the use of homeopathy in healthcare. For example, placebo treatments are not inevitably beneficial, and they can even endanger the health of patients. Even homeopathic remedies, containing no active ingredients, can carry risks. We will discuss the issue of safety in homeopathy and in relation to other alternative therapies at the end of the next chapter.

In the meantime, we will end this chapter by briefly considering another negative aspect of using placebo-based treatments such as homeopathy, namely the cost. This issue has been highlighted by Professor David Colquhoun, a pharmacologist who in 2006 criticized the sale of a homeopathic first-aid kit:

> All the 'remedies' in this kit are in the 30C dilution. They therefore contain no trace of the substance on the label. You pay £38.95 for a lot of sugar pills. To get even one molecule you'd have to swallow a sphere with a diameter equal to the distance from the Earth to the sun. That is hard to swallow.

If a person is going to spend £38.95 on a first-aid kit, then surely it is better to spend the money on real medicines that are genuinely effective, as opposed to wasting it on fake medicines, such as homeopathy, which offer only a placebo benefit. Perhaps the most extreme example of a homeopathic rip-off is a remedy called *Oscillococcinum*. The following paragraph, which is from an article published in the magazine *U.S. News and World Report* in 1996, underlines the utter absurdity and profiteering that underpins the homeopathic industry:

> Somewhere near Lyon, France, sometime this year, officials from the French pharmaceutical firm Boiron will slaughter a solitary duck and extract its heart and liver – not to appease the gods but to fight the flu. The organs will be used to make an over-the-counter flu medicine, called Oscillococcinum, that

will be sold around the world. In a monetary sense, this single French duck may be the most valuable animal on the planet, as an extract of its heart and liver form the sole 'active ingredient' in a flu remedy that is expected to generate sales of $20 million or more. (For duck parts, that easily beats out foie gras in terms of return on investment.) How can Boiron claim that one duck will benefit so many sick people? Because Oscillococcinum is a homeopathic remedy, meaning that its active ingredients are so diluted that they are virtually nonexistent in the final preparation.

In fact, the packaging boldly states that each gram of medication contains 0.85 grams of sucrose and 0.15 grams of lactose, which are both forms of sugar. In other words, Oscillococcinum is a self-declared 100 per cent sugar pill.

How can remedies free of active ingredients derived from a single duck be worth $20 million? This has to be the ultimate form of medical quackery.

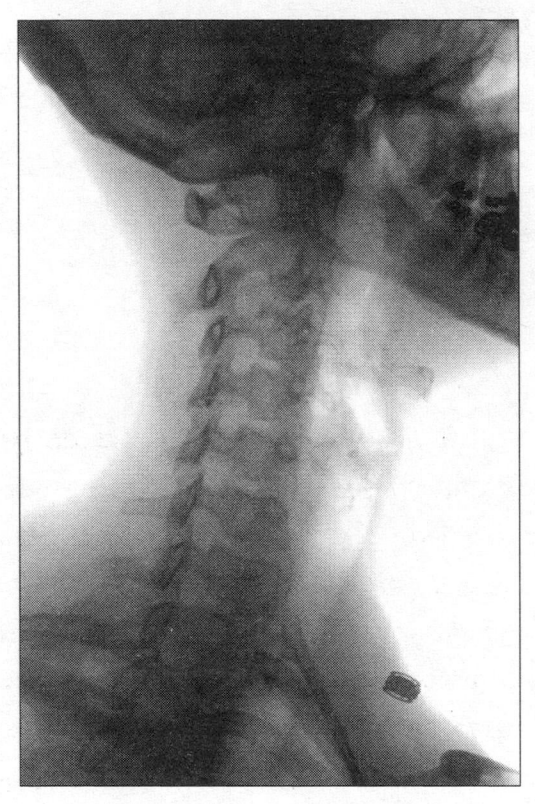

4 The Truth About Chiropractic Therapy

'... at the heart of science is an essential balance between two seemingly contradictory attitudes – an openness to new ideas, no matter how bizarre or counterintuitive, and the most ruthlessly sceptical scrutiny of all ideas, old and new. This is how deep truths are winnowed from deep nonsense.'

Carl Sagan

Chiropractic Therapy

A form of treatment developed at the end of the nineteenth century, which involves manual adjustments of the spine. Although some chiropractors focus on treating back pain, many others also treat a whole range of common illnesses, such as asthma. The underlying theory claims that manipulating the spine is medically beneficial because it can influence the rest of the body via the nervous system.

CHIROPRACTORS, WHO USUALLY DEAL WITH BACK OR NECK problems by manipulating the spine, are becoming such an established part of the healthcare system that many readers will be surprised to see chiropractic therapy included in a book about alternative medicine. After all, many conventional doctors refer their patients to chiropractors, and many insurance plans are willing to cover such treatments. This is particularly true in America, where chiropractors are most widespread, and where roughly $10 billion is spent annually on chiropractic treatment. As well as being an established part of the American healthcare system, chiropractors are becoming increasingly popular – between 1970 and 1990 their numbers tripled, and in 2002 there were 60,000 chiropractors practising in North America. It is expected that this figure will almost double by 2010, whereas the number of medical physicians will have increased by only 16 per cent.

Perhaps the most significant indication that chiropractors have become part of the medical mainstream is that they are licensed in all fifty US states, and they also have legal recognition in many other countries. For example, chiropractors in the United Kingdom are regulated by statute, which means that they have a similar standing to that of doctors and nurses. So, bearing all this in mind, why do chiropractors deserve to be labelled as alternative therapists?

The chiropractic approach to medicine emerged towards the end of the nineteenth century with a radically new view on health. The founders of chiropractic therapy argued that poor health was due to *subluxations*, by which they meant slight misalignments of the vertebrae in the spine. In turn, they believed that subluxations interfered with the flow of so-called *innate intelligence* (akin to a life force or vital energy), which then led to health problems of all sorts. But there is no evidence for the existence of innate intelligence or its role in health. The concepts of innate intelligence and subluxations are as mystical and as baffling as the concepts of Ch'i in acupuncture or extreme dilution in homeopathy, which means it makes no sense at all from a modern scientific point of view. That is why chiropractic treatment is still considered by many as an alternative medicine – despite its current popularity.

But, if we temporarily suspend disbelief and leave the underlying philosophy to one side, the key question is straight-forward: does chiropractic therapy help patients? Fortunately, this is a question that has been addressed thanks to evidence-based medicine and the use of clinical trials.

So far, evidence-based medicine has generated a pessimistic view of alternative medicine. Acupuncturists and homeopaths have spent centuries developing treatments to help patients, and yet scientists have examined the evidence, mainly from clinical trials, and concluded that these therapies are over-hyped. Acupuncture appears to be nothing more than a placebo for everything except some types of pain and nausea, and the jury is out even for these conditions. Worse still, homeopathic remedies have failed to be more effective than placebos in the treatment of every known condition.

Some readers may start to suspect that evidence-based medicine is somehow biased against alternative medicine. Perhaps acupuncture and homeopathy are actually valid therapies, and instead maybe it is in fact the clinical trial that is at fault? Perhaps the clinical trial is part of an establishment conspiracy cooked up by doctors and scientists to protect themselves from the interference of meddling outsiders? Just in case you are harbouring any such suspicions, let us take another look at the clinical trial and evidence-based medicine in general before examining the evidence for and against chiropractic therapy.

Evidence-based tea

The core principle of the trial is simple and can be traced back as far as the thirteenth century, when the Holy Roman Emperor Frederick II conducted an experiment to find out the effects of exercise on digestion. Two knights consumed identical meals, and then one went hunting while the other rested in bed. Several hours later, both knights were killed and the contents of their alimentary canals were examined. This revealed that digestion had progressed further in the sleeping knight. It was crucial to have two knights undergoing different levels of exercise, active and at rest, as it allowed the degree of digestion in one to be compared against the other. The key point of a trial is to compare the consequences of two or more situations.

The modern clinical trial, as developed by James Lind to test cures for scurvy in the eighteenth century, is less brutal than Frederick II's trial, but the central idea is the same. If, for example, a novel treatment is to be tested, then it needs to be compared against something else, known as the control. That is why the novel treatment is given to one group of patients and the control is given to another group. The control can be an established treatment, or a placebo or anything at all. Afterwards the patients in both groups are assessed, so that the effect of the novel treatment can be compared against the control.

Sir Ron Fisher, a British pioneer of the use of trials in the twentieth century, used to recount a story that amply demonstrated the simplicity and power of the trial. While at Cambridge, he became embroiled in an argument over how to make the ideal cup of tea. A woman insisted that it tasted worse if milk was added to the tea as compared to when tea was added to the milk, but the scientists around the table argued that it made no difference at all. Fisher immediately proposed a trial – in this case the comparison was the taste of milk added to tea versus the taste of tea added to milk.

Several cups were made with milk added to the tea, and several with tea added to the milk, and the woman was challenged to identify which was which. Although the cups of tea were prepared in secret and were identical in all other ways, the woman could indeed correctly recognize in each case whether the tea had been added to the milk or vice versa. The trial had shown that there was a difference, that the woman was right and that the scientists were wrong. In fact, there is a good scientific reason why the two forms of tea should taste different. Milk added to tea leads to a less satisfying cup, because the milk becomes superheated and this causes proteins in the milk to deteriorate – these proteins then taste slightly sour.

Fisher used this simple example as the basis for an entire book on scientific testing, *The Design of Experiments*, which went into great detail about the subtleties of trials.

Despite its sheer simplicity and powerful ability to get to the truth, some alternative therapists argue that the clinical trial is a harsh test, which is somehow biased against their treatments. But that sort of attitude betrays a skewed understanding of the clinical trial, which merely seeks to establish the truth, regardless of the type of treatment being examined. In fact, the clinical trial provides a wholly unbiased and truly fair test of any medical treatment, either conventional or alternative. The unbiased nature of the clinical trial is demonstrated by the fact that the history of mainstream medicine is littered with apparently good ideas from conventional doctors that clinical trials proved to be useless or harmful.

For example, Bill Silverman, an American paediatrician who died in 2004, was a committed advocate of the clinical trial, even though he realized that it was a double-edged sword, capable of either validating or crushing any treatment. In 1949 he began working at the newly opened premature-infant station at the Babies Hospital in New York, and within a few weeks he was dealing with a premature baby suffering from a problem known as retinopathy of prematurity (ROP), which can result in permanent blindness. The baby was the child of the hospital's biochemistry professor, whose wife had previously had six miscarriages. As this was the first time that the professor's wife had successfully given birth, Silverman was particularly distressed at the prospect of the child becoming blind. Grasping at straws, he decided to administer a newly discovered hormone known as ACTH (adrenocorticotropic hormone), which had not previously been used to treat newborn infants. Although it was a fairly hit-and-miss approach, with Silverman varying the dosage according to the baby's response, the end result was that she gained weight, her eyesight recovered and eventually she went home happy and healthy.

Inspired by this recovery, Silverman continued his ACTH treatment with subsequent cases of ROP. Furthermore, he compared his results with the recovery rates of babies with ROP at Lincoln Hospital, which was not offering ACTH treatment. The comparison was striking. Silverman gave ACTH to thirty-one babies suffering with ROP – twenty-five left with normal vision, two with near-normal vision, two with vision in just one eye and only two lost their sight completely. On the other hand, Lincoln Hospital had seven babies with ROP – they all lost their sight, except one.

For many doctors, the existing data set (thirty-one babies treated with ACTH with a success rate of 80 per cent versus seven untreated babies with a recovery rate of only 14 per cent) would seem convincing enough. It would have been easy for Silverman to have continued with this therapy and recommended it to colleagues as a method for preventing blindness, but instead he had the humility and courage to question his own discovery. In particular, Silverman could see that his pilot study fell short of the rigour demanded by a high-quality clinical trial. For example, the babies were not randomly assigned to the treatment or non-treatment groups, so maybe the babies at Lincoln Hospital were suffering from particularly serious problems, hence their lower recovery rate. Or maybe Lincoln Hospital's lack of success was a result of poorly trained staff or lack of equipment. Or maybe Lincoln Hospital was just unlucky – after all, the numbers involved were relatively small. To be confident about the efficacy of ACTH, Silverman decided to conduct a properly randomized controlled clinical trial.

Premature babies with ROP were randomly assigned to an ACTH treatment group or a no-treatment control group within the same hospital. Both groups were treated identically, apart from the use or not of ACTH. Within a few months the results emerged. An impressive 70 per cent of the infants treated with ACTH completely recovered their sight. Remarkably, the results in the control group were even more impressive, with an 80 per cent recovery rate. Babies in the untreated group had fared slightly better in terms of avoiding blindness, and moreover they suffered fewer fatalities compared to babies in the group treated with ACTH. It seemed that ACTH offered no benefit to babies and also had side-effects. A follow-up study confirmed the results of Silverman's rigorous clinical trial.

The initial results from the Lincoln Hospital were abnormally poor, which had fooled Silverman into believing that he had discovered a powerful new treatment, but he had been wise enough not to be complacent and rest on his laurels. Instead, he re-tested his own hypothesis and disproved it. Had he had not been so critical of his own work, subsequent generations of paediatricians might have followed his example and administered ACTH, a useless, expensive and potentially harmful treatment.

Silverman was a passionate believer in the randomized clinical trial as the tool for questioning and improving the care of babies, which made him an unusual figure among doctors in the 1950s. Although researchers were convinced of the importance of evidence in determining best practice, the doctors on the ground still tended to be overconfident about their gut instincts. They had faith in their own sense of what the ideal conditions should be for helping premature babies, but according to Silverman this was a primitive way of deciding serious health issues:

> Like the approach taken by farmers caring for newborn piglets, conditions considered ideal for survival were provided, and it was assumed that those who were 'meant' to survive would do so. But none of these purportedly 'ideal conditions' had ever been subjected to formal parallel-treatment trials . . . almost everything we were doing to care for premature infants was untested.

Doctors in the 1950s preferred to rely on what they had seen with their own eyes, and would typically respond to patients with the mantra 'in my experience'. It did not seem to matter to doctors that their personal experience might be limited or misremembered, as opposed to the evidence from research trials, which would be extensive and meticulously documented. That is why Silverman was determined to instil a more systematic approach among his colleagues, and he was supported in his mission by his former tutor Richard Day:

> Like Dick, I was completely sold on the numerical approach; soon we were making nuisances of ourselves by criticizing the

subjective 'in-my-experience' reasoning of our co-workers . . .
I was increasingly aware that the statistical approach was
anathema to free-wheeling doctors who resented any doubts
being expressed about the effectiveness of their untested treatments.

Half a century later, today's doctors are much more
accustomed to the concept of evidence-based medicine, and
most accept that a well-designed randomized clinical trial is
crucial for deciding what works and what does not. The
purpose of this book has simply been to apply these same
principles to alternative medicine. So what does evidence-
based medicine say about chiropractic therapy?

Manipulating patients

When patients visit a chiropractor, they are usually suffering
from back or neck pain. After taking a medical history, the
chiropractor will embark on a thorough examination of
the back, particularly the bones of the spine, called *vertebrae*.
This will include looking at the patient's posture and overall
mobility, as well as feeling along the spine to assess the
symmetry and mobility of each spinal joint. Often X-ray
images or sometimes MRI scans are also used to give a
detailed view of the vertebrae. Any misalignment in the spine
is then corrected in order to restore the patient's health.
Chiropractors see the spinal column as a complex entity, such
that each vertebra affects all the others. Hence, a chiropractor
might work on a patient's upper spine or neck in order to treat
pain in the lower back.

The hallmark treatment of the chiropractor is a range of
techniques known as *spinal manipulation*, which is intended to
realign the spine in order to restore the mobility of joints.
Chiropractors also call this an *adjustment*. It can be a fairly

aggressive technique, which pushes the joint slightly beyond what it is ordinarily capable of achieving. One way to think about spinal manipulation is as the third of three levels of flexibility that can be achieved by a joint. The first level of flexibility is that which is possible with only voluntary movement. A second and higher level of flexibility can be achieved by exerting an external force, which pushes the joint until there is resistance. The third level of flexibility, which corresponds to spinal manipulation, involves a thrusting force that pushes the joint even further. The chiropractor will submit the vertebrae of the spine to this third level of motion by using a technique called *high-velocity, low-amplitude thrust*. This means that the chiropractor exerts a relatively strong force in order to move the joint at speed, but the extent of the motion needs to be limited, because otherwise there would be damage to the joint and its surrounding structures. Although spinal manipulation is often associated with a cracking sound, this is not a result of the bones crunching against each other or a sign that bones are being put back in their right place. Instead, the noise is caused by the release and popping of gas bubbles, which are generated when the fluid in the joint space is put under severe stress.

If you have never visited a chiropractor, then the easiest way to imagine spinal manipulation is by analogy to an experiment you can do with your hand. Position your right forearm vertically upwards and hold your right hand flat, with the palm facing up – as if you are carrying a tray of drinks. Your wrist should be able to bend backwards so far that your flat hand begins to dip slightly below the horizontal – this is what we have called level-one flexibility. If you use your left hand to press steadily and firmly downwards on your right palm, then the wrist can be bent a little further down by a few degrees, which is level-two flexibility. Imagine – *and please do not do this* – that your left hand applied an additional short rapid thrust on your right hand, thus bending it down even further by a small amount. This would be level-three flexibility, akin to the sort of action involved in spinal manipulation via a high-velocity, low-amplitude thrust.

Because spinal manipulation is the technique that generally distinguishes chiropractors from other health professionals, it has been at the centre of efforts to establish the medical value of chiropractic therapy. Researchers have conducted dozens of clinical trials in order to evaluate spinal manipulation, but they have tended to generate conflicting results and have often been poorly designed. Fortunately, as with acupuncture and home-opathy, there have been several systematic reviews of these trials, in which experts have attempted to set aside the poor trials, focus on the best-quality trials and establish an overall conclusion that is reliable.

In fact, there have been so many systematic reviews that in 2006 Edzard Ernst and Peter Canter at Exeter University decided to take all of the current ones into account in order to arrive at the most up-to-date and accurate evaluation of chiropractic therapy. Published in the *Journal of the Royal Society of Medicine*, their paper was entitled 'A systematic review of systematic reviews of spinal manipulation'. Ernst and Canter's review of recent reviews covered spinal manipulation in the context of a large range of conditions, but for the time being we will concentrate on the most common problems dealt with by chiropractors, namely back and neck pain. In this context they took into account three reviews looking at back pain alone, two reviews looking at neck pain alone and one review that covered both neck and back pain.

The individual reviews came to varying conclusions. In the case of neck pain, two reviews concluded that spinal manipulation was ineffective, although one of them did find some evidence that chiropractic manipulation could be effective when used in combination with standard treatments. However, the combination effect is hard to disentangle, so it would be difficult to draw anything significant from this. The third review was more positive, concluding that spinal manipulation offered patients a moderate benefit, but it is worth noting that the lead author on this review was a chiropractor. Ernst and Canter had previously shown that chiropractors tend to generate more optimistic conclusions than scientists, perhaps because they have an emotional investment in the result. All in all, the evidence was insubstantial.

For acute back pain, there was more of a consensus that spinal manipulation could be effective. Each review suggested that, on average, patients benefitted from the sort of treatments offered by chiropractors, but there was disagreement over the extent of that benefit and the evidence was not conclusive. The fact that chiropractic spinal manipulation might help with back pain is not a major milestone in the history of medicine – but it is particularly noteworthy in the context of this book, because this is the most significant evidence so far that an alternative treatment might genuinely help patients.

On the other hand, this conclusion should not be interpreted as an endorsement of chiropractors or a recommendation that patients with back pain should try spinal manipulation. The key question is not merely 'does spinal manipulation work?', but rather 'does spinal manipulation work better than other forms of treatment?'

Dealing with back problems is notoriously difficult, and conventional medicine has struggled to develop truly effective treatments. In terms of dealing with the underlying problem, doctors might recommend physiotherapy or exercise. And in terms of dealing with the symptoms, doctors often prescribe non-steroidal anti-inflammatory drugs (NSAIDs), such as ibuprofen. These approaches are, however, only mildly or marginally effective. A truly life-changing cure for back pain has not yet been found.

When the two approaches are compared against each other, spinal manipulation versus conventional medicine, the result is that each is just about as effective (or ineffective) as the other. Indeed, this was one of the main conclusions of Ernst and Canter's review of reviews: spinal manipulation might help those who suffer with back pain, but conventional approaches offer similarly marginal levels of benefit.

In a situation where two or more rival treatments match each other in terms of their effectiveness, there are several other deciding factors that determine which one is best. The simplest determining factor is often cost, which mitigates strongly against chiropractors, who generally charge a great deal for their services based on the misguided claim that their treatment is superior to conventional treatments. Compare ten sessions with a chiropractor at £50 each with regular exercise or ibuprofen, which are both relatively cheap, and the price difference becomes obvious.

Furthermore, there are more important factors which also favour conventional treatment over chiropractic spinal manipulation. In fact, there are serious problems with chiropractic philosophy and practice, both of which should raise major concerns for prospective patients. These issues are closely linked to the early development of this form of treatment, so in order to appreciate them properly we will take a historical detour and explore the origins of chiropractic therapy.

The bone-setting panacea

The first documented account of manipulating the spine for treating people dates back to Hippocrates in around 400 BC. In order to deal with back problems, he asked patients to lie face down on a board and his assistants applied traction by pulling on the head and feet. At the same time, Hippocrates pressed on the painful part of the spine, or sat on it, or bounced up and down, or walked along it. We do not, by the way, recommend you try this at home!

As the centuries passed, it became the responsibility of specialists known as *bone-setters* to treat bones that were broken, misaligned or dislocated. In Norway the local bone-setter was often a first-born child, whereas in Ireland the bone-setter was typically the seventh-born, but birth order did not matter in Scotland as long as the person had been born feet first. Because bone-setters were not usually formally educated and were not therefore part of the medical establishment, they often drew criticism from physicians. For example, Sarah Mapp, who was one of the most famous bone-setters in London in the 1730s, was nicknamed 'Crazy Sally' by many physicians. Percival Pott, an eminent English surgeon who was the first to demonstrate that soot could cause cancer in chimneysweeps, went further and called her an 'ignorant, illiberal, drunken, female savage'. On the other hand, Sir Hans Sloane, who was President of the Royal College of Physicians, had sufficient respect for 'Crazy Sally' to ask her to treat his niece's back injury.

Chiropractic therapy, which emerged out of the bone-setting tradition, was founded by Daniel David Palmer, who was born near Toronto, Canada, in 1845 and who moved to Iowa at the age of twenty. Palmer gradually developed an interest in medicine, which included spiritual and magnetic healing, but his interest in the potential of spinal manipulation can be traced to a specific event that took place on 18 September 1895. Here is how Palmer later recorded the event:

Harvey Lillard, a janitor in the Ryan Block, where I had my office, had been so deaf for 17 years that he could not hear the racket of a wagon on the street or the ticking of a watch. I made enquiry as to the cause of his deafness and was informed that when he was exerting himself in a cramped, stooping position, he felt something give in his back and immediately became deaf. An examination showed a vertebra racked from its normal position. I reasoned that if the vertebra was replaced, the man's hearing should be restored. With this object in view, a half hour talk persuaded Mr Lillard to allow me to replace it. I racked it into position by using the spinous process as a lever and soon the man could hear as before.

On its own, this incident would not have started a revolution, but Palmer treated a second patient in a similar manner:

Shortly after this relief from deafness, I had a case of heart trouble which was not improving. I examined the spine and found a displaced vertebra pressing against the nerves which innervate the heart. I adjusted the vertebra and gave immediate relief . . . Then I began to reason if two diseases, so dissimilar as deafness and heart trouble, came from impingement, a pressure on nerves, were not other diseases due to a similar cause? Thus the science (knowledge) and art (adjusting) of Chiropractic were formed at that time. I then began a systematic investigation for the cause of all diseases and have been amply rewarded.

Palmer believed that he had stumbled upon a new medical technique. He was so convinced that chiropractic therapy offered a novel approach to healthcare that he opened the Palmer School of Chiropractic in 1897 in Davenport, Iowa. His reputation and charisma rapidly attracted many students to the school, where the main teaching resource was a textbook entitled *The Chiropractor's Adjuster*, written by Palmer himself. This outlined every detail of his chiropractic therapy in its 1,000 pages, including how Palmer came to name his new treatment: 'Rev. Samuel H. Weed of Portland selected for me at my request two Greek words, *cheir* and *praxis*, meaning when combined "done by hand", from which I coined the word "chiro-practic".'

Perhaps the most surprising feature of Palmer's chiropractic therapy was its ambition. Having allegedly treated deafness and a heart condition by realigning the spine of his patients, he was confident that spinal manipulation could deal with all the ills of the human race. For Palmer, chiropractic

Daniel David Palmer

therapy was not primarily about treating back problems. He explicitly wrote: 'Ninety-five per cent of all diseases are caused by displaced vertebrae.'

This statement might seem shocking to us, but it made perfect sense to Palmer, who viewed the spine as key to the health of the entire body. He was keenly aware that the spine provides the highway that connects the brain and the spinal cord to the rest of the body by way of the peripheral nervous system. Hence, according to Palmer, displaced vertebrae would impact on particular neural pathways, negatively influence the organs connected via this pathway and thereby cause diseases. Consequently, if chiropractors realigned these displaced vertebrae then they could cure diseases: not just deafness and heart disease, but also everything from measles to sexual dysfunction.

This is already an extraordinary claim, and it appears even more bizarre when phrased in Palmer's own language. As mentioned earlier in the chapter, Palmer used the term

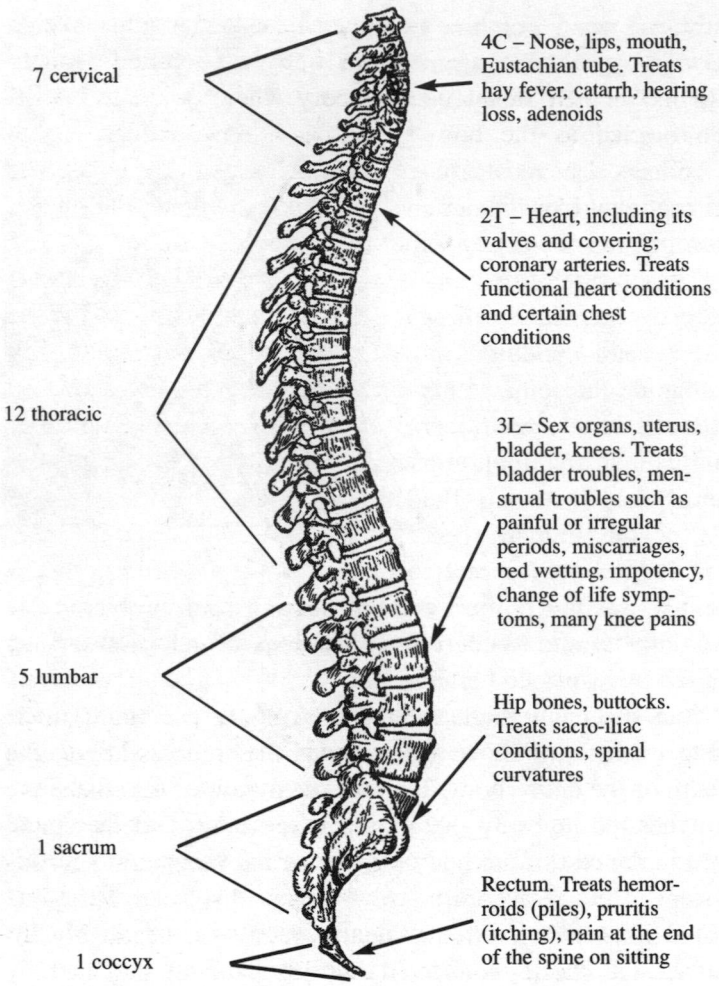

7 cervical

4C – Nose, lips, mouth, Eustachian tube. Treats hay fever, catarrh, hearing loss, adenoids

2T – Heart, including its valves and covering; coronary arteries. Treats functional heart conditions and certain chest conditions

12 thoracic

3L– Sex organs, uterus, bladder, knees. Treats bladder troubles, menstrual troubles such as painful or irregular periods, miscarriages, bed wetting, impotency, change of life symptoms, many knee pains

5 lumbar

Hip bones, buttocks. Treats sacro-iliac conditions, spinal curvatures

1 sacrum

Rectum. Treats hemorrhoids (piles), pruritis (itching), pain at the end of the spine on sitting

1 coccyx

Figure 5

A chiropractic chart shows how each vertebra relates to different parts of the body and is responsible for various ailments. This simplified chart shows the ailments corresponding to only some of the vertebra. For example, a misaligned third lumber vertebra could cause bladder problems, and realignment could cure this. When Palmer cured his first two patients, he presumably manipulated the fourth cervical vertebra and the second thoracic vertebra, as these are linked with hearing loss and heart problems.

'subluxation' to describe a displacement in the spine, which resulted in a blockage of the body's so-called 'innate intelligence'. He developed a theory whereby innate intelligence acted as the body's guiding energy, carrying both metaphysical and physiological significance. This is why he believed that blocking its flow seriously disrupted the body's harmony and could lead to all manner of diseases.

It is important to stress that the term 'innate intelligence' is utterly meaningless beyond Palmer's unique view of the human body. On the other hand, the term 'subluxation' is used in orthodox medicine, but has a meaning that has nothing to do with blocking innate intelligence. If a doctor talks about 'subluxation', it simply means a partial dislocation of any joint, such as a twisted ankle. In short, Palmer's 'innate intelligence' and 'subluxation' carry no scientific significance.

The concept of innate intelligence was so strange that it seemed as if chiropractic therapy was not only a new medical doctrine, but also a new religion. Indeed, Palmer viewed God as the Universal Intelligence, guiding the totality of existence, which meant that innate intelligence represented God's guiding influence within the human body. In Palmer's own words, 'I am the founder of chiropractic in its science, in its art, in its philosophy and in its religious phase'. He even likened himself to 'Christ, Mohamed, Jo. Smith [founder of the Latter Day Saints movement], Mrs. Eddy [founder of the Christian Science church], Martin Luther and other[s] who have founded religions.'

Conventional doctors were suspicious of Palmer's quasi-religious philosophy, and they were particularly angered by his extraordinary claim that the spine was the root cause of disease and that spinal manipulation was the best way to cure patients. They were annoyed by his boast that 'chiropractic is a science of healing without drugs', and they were alarmed by his refusal to acknowledge the role of germs in causing many of the diseases prevalent at the time. Not surprisingly, it was not long before there was a campaign against Palmer, led by a local doctor named Heinrich Matthey. He accused Palmer of teaching an unproven medical concept and practising medicine without a licence. In fact, this led to Palmer being taken to court three times and on the third occasion, in 1906, he was sentenced to time in jail when he refused to pay a fine. If anything, this strengthened the fast-growing movement: chiropractic therapy had its first martyr, and many more would follow.

D. D. Palmer's son was Bartlett Joshua Palmer, and it was he who continued to promote chiropractic therapy while his father was indisposed. He became successful in his own right, so much so that he was able to buy the first car in Davenport, but unfortunately in 1913 he ran over his father at the Palmer School of Chiropractic homecoming parade. D. D. Palmer died just a few weeks later – officially the cause of the death was recorded as typhoid, but it seems more likely that his death was a direct result of injuries caused by his son. Indeed, there is speculation that this was not an accident, but rather a case of patricide. Father and son had become bitter rivals over the leadership of the chiropractic movement. Also B. J. Palmer had always resented his father and the way that he had treated his family:

> When each of our sisters reached eighteen, they were driven out of home and onto the streets of Davenport to make their living any way they could . . . All three of us got beatings with straps until we carried welts, for which father was often arrested and spent nights in jail . . . Father was so deeply involved and so busy with thinking and writings on Chiropractic, he hardly knew he had any children.

B. J. Palmer, who already led the Universal Chiropractic Association, became the new undisputed figurehead of the movement. He was a smart operator and shrewd entrepreneur. He rapidly accumulated a large fortune by teaching students and treating patients. On top of all this, in 1924 he started a lucrative sideline in leasing *neurocalometers*, which supposedly helped chiropractors by detecting subluxations. Palmer was very proud of his invention and promoted it widely, but from a modern point of view we can see that it was clearly a worthless piece of technology. The neurocalometer contained nothing more than a simple *thermocouple*, which is a standard piece of electrical equipment designed to measure temperature. Hence, it would have been useless for detecting misalignments in the spine or pinched nerves. Although each neurocalometer cost less than $100 to build, he initially leased them for ten years for $1,150, and then he increased the price to $2,200. To put this figure into context, $2,200 would have been enough to buy a house in Iowa in the 1920s, and yet Palmer persuaded over 2,000 graduates from his college and other chiropractors to buy his bogus invention.

Not surprisingly, his customers ended up dissatisfied. An attorney acting on behalf of one of the disgruntled customers attempted to sue Palmer: 'In all our experience as practicing attorneys, nothing more closely resembling a fraud and a swindle has ever been brought to our personal attention than this proposition which your school is submitting to its graduates.'

In such situations, Palmer would repair his reputation by promoting himself on WOC, one of America's pioneering radio stations, which he had established in 1922. Although it carried programmes on a range of subjects, such as current affairs and cookery, it also broadcast lectures by Palmer as well as other programmes directly related to chiropractic therapy. Its audience stretched across large parts of America and Canada, and Palmer even claimed he had listeners in Scotland, Samoa and at the North Pole.

Thanks to his radio station and other clever marketing techniques, Palmer oversaw the growth of the chiropractic movement over the next few decades, not just in America but also in Europe. For example, the British Chiropractic Association was founded in 1925 and the European Chiropractors' Union formed in 1932, by which time there were 126 chiropractors in Britain, seventy-six across Norway, Denmark and Sweden, and a few dozen others in Ireland, Belgium and elsewhere.

Meanwhile, back in America, chiropractors were coming under increasing pressure from the medical establishment, which disapproved of their philosophy and methods. Doctors continued to encourage the arrest of chiropractors for practising medicine without a licence, and by 1940 there had been over 15,000 prosecutions. Palmer strongly endorsed the Universal Chiropractic Association's policy of covering legal expenses and supporting members who had been arrested, which resulted in 80 per cent of chiropractors walking free from court.

When the legal route failed to dampen chiropractic spirits, the American Medical Association (AMA) tried other tactics, which culminated in 1963 in the formation of the Committee on Quackery. Its secretary, H. Doyl Taylor, wrote a memo to the AMA Board of Trustees, which reaffirmed that the Committee on Quackery considered its prime mission to be the 'containment of Chiropractic and ultimately the elimination of Chiropractic.' The Committee's activities included lobbying to keep chiropractors outside the Medicare health-insurance programme, and arguing that chiropractic therapy should not be recognized by the US Office of Education.

This antagonism might seem unreasonable, but remember that the medical establishment had several reasons for despising chiropractors. These included their belief in the unscientific notion of innate intelligence, their denial that bacteria and viruses cause many diseases, and their conviction that realigning a patient's spine could cure every ailment. On top of all this, conventional doctors were shocked by the fact that many chiropractors were fond of the *E-meter*, another bizarre diagnostic gadget. Invented in the 1940s by a chiropractor named Volney Mathison, the E-meter has a needle that swings back and forth across a scale when a patient holds on to two electrical contacts – apparently this is enough to determine a patient's state of health. The E-meter was also widely used by the Church of Scientology, so much so that many Scientologists believe that it was invented by their founder L. Ron Hubbard. Unfortunately, the E-meter is nothing more than a piece of technical hocus-pocus, which is why in 1963 the US Food & Drug Administration seized more than 100 of them from the Founding Church of Scientology. In many ways, the E-meter bears a resemblance to the equally bogus neurocalometer, invented two decades earlier by B. J. Palmer.

Conventional doctors were equally dismissive of *applied kinesiology*, a method invented in 1964 by a chiropractor called George J. Goodheart, who argued that diseases could be identified by manually testing the strength of muscle groups. A patient's muscles supposedly become immediately stronger if a treatment is beneficial, or the muscles become weaker if the treatment is harmful, or if a toxin or allergen is brought close to the body. Typically, the patient holds out an arm and a tester pushes against it to feel the strength and steadiness of the resistance. This is, of course, a highly subjective measurement, and it is hard to imagine why it should have any medical value. Indeed, controlled trials show that the claims of applied kinesiology have no basis in reality.

As far as the AMA was concerned, all these problems were compounded by the ambition of many chiropractors to act as primary care givers. In other words, chiropractors argued that they could replace general practitioners because they could also offer regular check-ups, long-term preventative treatments and cures for many conditions. In the 1950s and 1960s it was possible to find adverts for chiropractors with claims such as 'there are very few diseases, as they are understood today, which are not treatable by Chiropractic method', or 'Correction and treatment of both acute and chronic polio by chiropractic methods have been unusually successful.'

The AMA continued to fight back with its concerted effort to eliminate the chiropractic profession, but in 1976 its campaign suddenly backfired. 'Sore Throat', an anonymous source within the AMA, leaked material that revealed the details and the extent of the AMA's campaign, which prompted Chester A. Wilk, a chiropractor from Chicago, to file an anti-trust lawsuit against the AMA. Wilk was arguing that the AMA's campaign against chiropractors amounted to anti-competitive behaviour, and that the medical establishment was merely trying to corner the market in treating patients.

After dragging on for over a decade, the lawsuit eventually ended in 1987. Judge Susan Getzendanner, who had presided over the case, ruled that the AMA had indeed acted unfairly against chiropractors:

Evidence at the trial showed that the defendants took active steps, often covert, to undermine chiropractic educational institutions, conceal evidence of the usefulness of chiropractic care, undercut insurance programs for patients of chiropractors, subvert government inquiries into the efficacy of chiropractic, engage in a massive disinformation campaign to discredit and destabilize the chiropractic profession and engage in numerous other activities to maintain a medical physician monopoly over health care in this country.

The AMA took the decision to the Supreme Court, but the appeal failed in 1990 and thereafter the AMA was forced to alter its attitude. For example, it could no longer discourage its members from collaborating with chiropractors. Although the medical establishment had fought against this move, it had to acknowledge that it resulted in two undoubtedly positive outcomes. First, those doctors who collaborated with chiropractors persuaded many of them to be more sympathetic to the ideas of conventional medicine. Second, it also encouraged many chiropractors to rethink their attitude to their own chiropractic therapy. In fact, many chiropractors were already becoming increasingly disillusioned with the outlandish claims of their founding fathers. Although these practitioners were still committed to using chiropractic therapy to treat musculoskeletal problems, they were reluctant to treat other conditions and were suspicious about the concept of innate intelligence. In short, these rebel chiropractors adopted a more defined job description, namely back specialists. They became known as *mixers*, because they were willing to mix traditional chiropractic therapy with elements of mainstream medicine.

By contrast, chiropractors who strictly adhered to Palmer's philosophy were known as *straights*. They firmly believed every word Palmer had preached, including his core belief that a perfectly aligned spine would guarantee the flow of the 'innate' and thus promote wellbeing throughout the entire body. The split between straights and mixers soon became bitter, with straights accusing mixers of betraying the chiropractic movement, and mixers accusing straights of being quacks. In 1998 Lon Morgan, a mixer, openly expressed his antagonism towards straight chiropractors and their odd beliefs: 'Innate Intelligence clearly has its origins in borrowed mystical and occult practices of a bygone era. It remains untestable and unverifiable and has an unacceptably high penalty/benefit ratio for the chiropractic profession.' Similarly, according to Joseph C. Keating, a mixer and a chiropractic historian: 'So long as we propound the "One cause, one cure" rhetoric of Innate, we should expect to be met by ridicule from the wider health science community.' In response, straights have accused mixers of not being real chiropractors, because they do not accept Palmer's basis for chiropractic therapy.

It is relatively easy to find out who is right – straights or mixers – because the former set of chiropractors claims that spinal manipulation cures everything and the latter tends to restrict its ambitions to the back and neck. Ordeal by clinical trial is the obvious method for settling such an argument. In fact, many clinical trials have been conducted to test the impact of spinal manipulation on a range of conditions, and many of these were covered within the review of reviews by Ernst and Canter, which was discussed earlier in this chapter. We have already considered their conclusions relating to back and neck pain, but now it is time to look at their other conclusions.

Ernst and Canter looked at ten systematic reviews based on seventy trials that considered spinal manipulation as a treatment for headaches, period pains, infantile colic, asthma and allergies. Their conclusions were universally negative – there was no evidence to suggest that chiropractors could treat any of these conditions.

This should not be very surprising, as there is no logical, rational or scientific reason why manipulating a patient's spine should treat, for example, allergies. Moreover, there is no evidence that a misaligned spine can cause any of these non-musculoskeletal conditions in the first place. Indeed, if spinal misalignments caused disease, then we would expect people with back pain to be more likely to suffer with other ailments, but in 1995 Donald Nansel and Mark Szlazak at the Palmer College of Chiropractic found no sign of this in the vast body of published medical literature: 'There is not the slightest suggestion that patients suffering from severe primary mechanical low back pain, for instance, are more prone to develop higher incidences of prostate or testicular carcinoma, colitis, ovarian cysts, endometriosis, pancreatitis, appendicitis, diabetes mellitus, or any other category of regionally or segmentally related organ disease.' In a follow-up study published two years later, the same researchers also failed to find any evidence that these diseases were more likely in 'patients with broken necks or broken backs, or patients with entire hips or shoulders blown apart by shotgun blasts'.

Although Ernst and Canter's review of reviews does not cover the impact of chiropractic manipulation on every non-musculoskeletal condition, it would be reasonable to conclude that chiropractors can offer nothing to help patients suffering from non-musculoskeletal conditions in general. This is partly because chiropractic therapy has failed whenever it has been tested as a treatment for specific non-musculoskeletal conditions, and it is partly because – and this is worth stressing again – there is no reason why spinal manipulation should help with conditions ranging from ear infections to irritable bowel syndrome.

Bearing all this in mind, the scientific evidence shows that it would be unwise to visit a chiropractor for anything other than a problem directly related to your back.

This might seem obvious, but several surveys suggest that between 11 per cent and 19 per cent of American chiropractic patients suffer from non-musculoskeletal conditions. These patients are attracted to these pointless treatments by practitioners who are willing to offer them. According to one survey, 90 per cent of American chiropractors think that the therapy should not be limited to musculoskeletal conditions, and another survey suggests that 78 per cent of Canadian chiropractors share this opinion – this indicates that the majority of North American chiropractors have straight tendencies. The percentages in Europe may be similar, particularly as supposedly responsible chiropractic bodies in European countries offer misleading information about the power of chiropractic therapy. For example, the General Chiropractic Council, which oversees chiropractic therapy in Britain, publishes a leaflet entitled 'What can I expect when I see a chiropractor?', which states that chiropractic therapy can lead to an improvement in 'some types of asthma, headaches, including migraine, and infant colic'. Yet it is well known that the evidence from trials fails to support these claims.

Some words of caution for patients

In short, the scientific evidence suggests that it is only worth seeing a chiropractor if you have a back problem. However, it is still important to be cautious. In particular, we will offer six pieces of advice that should be useful for anybody considering a visit to a chiropractor:

1   Make sure that your chiropractor is a mixer and not a straight. It would be unwise to be treated by a chiropractic fundamentalist, namely someone who believes in subluxations, innate intelligence and the ability of spinal manipulation to cure all diseases. The terms 'straight' and 'mixer' do not generally appear on a chiropractor's business card, so the best way to identify a straight is to ask about the range of conditions that he or she claims to treat – a straight chiropractor will offer treatments for respiratory conditions, digestive disorders, menstrual problems, ear infections, pregnancy-related conditions, infectious and parasitic conditions, dermatological diseases, acute urinary conditions and many other ailments.

2   If you visit a chiropractor and the problem is not resolved within six sessions, or there is no ongoing significant improvement within six sessions, then be prepared to

stop treatment and consult your doctor for advice. Chiropractors have a reputation for lengthy and expensive treatments, as demonstrated by a survey in 2006 that monitored ninety-six patients with acute neck pain. Although the patients generally reported improvements, the treatments required twenty-four visits on average, and in two cases there were more than eighty treatment sessions. It is likely that the majority of these recoveries had little to do with the chiropractic intervention, but were largely the result of time and the body's own natural healing processes.

3   Do not allow a chiropractor to become your primary healthcare provider, which might include preventative and maintenance treatments covering all health issues. In 1995, a survey showed that 90 per cent of American chiropractors considered themselves primary care providers, but they are rarely qualified to take on this role. Patients are often impressed by the fact that many chiropractors carry the title Doctor, but this does not mean that they have attended medical school. The title generally indicates Doctor of Chiropractic (DC), which merely means that a practitioner has completed a chiropractic course lasting four years.

4   Avoid chiropractors who rely on unorthodox techniques for diagnosing patients, such as applied kinesiology and the E-meter, which were described earlier. Such techniques are usually employed by straight chiropractors.

5   Check the reputation of your chiropractor before embarking on any treatment, because chiropractors are more likely than medical doctors to be involved in malpractice. According to a survey conducted in California in 2004, chiropractors were twice as likely as

medical doctors to be the subject of disciplinary actions. Even more worrying, the incidence rate for fraud was nine times higher for chiropractors than for doctors, and the rate for sexual boundary transgressions was three times higher for chiropractors than doctors.

6   Last, but not least, try conventional treatments before turning to a chiropractor for back pain. They are generally cheaper than spinal manipulation and just as likely to be effective. There are also other reasons for following the conventional route, but we will come to these later in the chapter.

The advice above is based on serious and well-founded criticisms of some elements of the chiropractic community. For example, chiropractors, particularly in America, have earned a reputation for zealously recruiting and unnecessarily treating patients. Practice-building seminars are commonplace and there are numerous publications aimed at helping chiropractors find and retain patients. In many cases the emphasis seems to be placed on economics rather than healthcare: the chiropractor Peter Fernandez is the author of a five-volume series called *Secrets of a Practice-Building Consultant*, which starts with a volume boldly titled *1,001 Ways to Attract Patients* and ends with *How to Become a Million Dollar a Year Practitioner*.

Many chiropractors are embarrassed by the obsessive profiteering of their colleagues. For instance, G. Douglas Anderson, writing in *Dynamic Chiropractic*, has argued that the chiropractic movement needs a radical overhaul:

It is high time we admit there is nothing conservative, holistic or natural about endless care, creating addiction to manipulation, or making unsubstantiated, cure-all claims. On the contrary, an excellent argument can be made that the variety of tricks, techniques and claims still used by a large percentage of our profession to keep fully functional, asymptomatic people returning for care is fraudulent.

According to Joseph C. Keating, who taught many chiropractors, the tendency to profiteer and mislead can be traced back to the founders of the chiropractic therapy, particularly B. J. Palmer: 'Indeed, the profession, as a unified body politic, has never truly renounced the marketing and advertising excesses modelled by B. J. and many clinical procedures and innovations since are noteworthy for the extraordinary and unsubstantiated claims which are made for them.' It seems that chiropractors are fond of manipulating their patients in both senses of the word.

Stephen Barrett, a US psychiatrist and medical writer, has been at the forefront of exposing other shady aspects of chiropractic therapy. For example, he conducted an experiment to see how four chiropractors would deal with the same healthy patient, a twenty-nine-year-old woman:

> The first diagnosed 'atlas subluxation' and predicted 'paralysis in fifteen years' if this problem was not treated. The second found many vertebrae 'out of alignment' and one hip 'higher' than the other. The third said that the woman's neck was 'tight'. The fourth said that misaligned vertebrae indicated the presence of 'stomach problems'. All four recommended spinal adjustment on a regular basis, beginning with a frequency of twice a week. Three gave adjustments without warning – one of which was so forceful that it produced dizziness and a headache that lasted several hours.

Barrett's study of chiropractors was neither exhaustive nor definitive, but his limited sampling did suggest that there is something rotten at the heart of the chiropractic profession. Chiropractors dealing with the same healthy individual could agree on neither the diagnosis nor which part of the spine was problematic – all that they could agree on was that regular chiropractic therapy was the solution. Perhaps this should not be surprising when we bear in mind that the underlying principles of chiropractic therapy, the notions of subluxations and innate intelligence, are meaningless.

In addition to all this, and even more worrying, is Barrett's last sentence, which mentions that his undercover patient suffered 'dizziness and a headache that lasted several hours'. This raises an important issue that we have not yet discussed, namely safety. Every medical treatment should offer the likelihood of benefit, but it will also, almost inevitably, carry a likelihood of side-effects. The key issue for patients is simple: does the likely extent of the benefit outweigh the likely extent of the adverse side-effects, and how does this risk–benefit analysis compare to other treatments? As we shall discuss below, the dangers of chiropractic therapy can be serious and in some cases life-threatening.

The dangers of chiropractic therapy

Often the first hazard encountered when visiting a chiropractor is undergoing an X-ray examination, which seems to be a routine procedure among many practitioners. A survey conducted across Europe in 1994 revealed that 64 per cent of patients received X-rays when visiting a chiropractor, and a survey of members of the American Chiropractic Association conducted in the same year suggested that 96 per cent of new patients and 80 per cent of returning patients were X-rayed. Although many chiropractic publications explicitly advise against the routine use of X-rays, these surveys reveal an almost cavalier approach to a technology that does carry a risk of causing cancer.

It is estimated that on average medical X-rays are responsible for 14 per cent of our annual exposure to radiation. Much of the remaining 86 per cent comes from natural sources such as radon gas seeping up through the ground. The increased risk of cancer due to X-rays is small, but it is not negligible. According to a paper published in the *Lancet* in 2004, roughly 700 of the 124,000 new cancer cases diagnosed each year in the UK are due to medical X-rays. Although X-rays therefore account for 0.6 per cent of new cancer cases, they continue to be used widely in medicine because they offer tremendous benefits in terms of diagnosing and monitoring patients. In other words, conventional doctors are willing to use X-rays because the benefits outweigh the potential harm, but at the same time they minimize the use of X-rays, employing them only when there is a clear reason to do so.

In contrast, chiropractors may X-ray the same patient several times a year, even though there is no clear evidence that X-rays will help the therapist treat the patient. X-rays can reveal neither the subluxations nor the innate intelligence associated with chiropractic philosophy, because they do not exist.

There is no conceivable reason at all why X-raying the spine should help a straight chiropractor treat an ear infection, asthma or period pains. Most worrying of all, many chiropractors require a full spine X-ray, which delivers a significantly higher dose of radiation than most other X-ray procedures.

This raises the question of why so many chiropractors are so keen to X-ray their patients. Partly they are blindly following a corrupt methodology and a bankrupt philosophy that has been passed down through the decades, while ignoring the latest advice from experts. On top of this, it is important to remember that X-raying patients is a very lucrative part of any chiropractic business.

In addition to the risk associated with X-rays, the actual manipulation of the spine can also have negative repercussions. In 2001, a systematic review of five studies revealed that roughly half of all chiropractic patients experience temporary adverse effects, such as pain, numbness, stiffness, dizziness and headaches. These are relatively minor adverse effects, but the frequency is very high, and this has to be weighed against the limited benefit offered by chiropractic therapy.

More worryingly, patients can also suffer serious problems such as dislocations and fractures. These hazards are more likely and more dangerous for older patients, who may be suffering from osteoporosis. For instance, in 1992 the *Journal of Manipulative and Physiological Therapeutics* reported the case of a seventy-two-year-old woman who visited a chiropractor complaining of back pain. She received twenty-three treatment sessions over the course of six weeks, which resulted in multiple spinal compression fractures.

On top of all these risks, there is an even more serious hazard associated with chiropractic therapy. To appreciate this hazard, we need to refer back to Figure 5 on page 196, which shows the structure of the spine. It consists of five regions – the coccygeal region is at the base, followed by the sacral, lumbar and thoracic regions, with the cervical region at the top. The most severe risk relates to manipulation of the cervical region. There are seven vertebrae that make up this region, running from the base of the neck to the back of the skull. This is one of the most flexible parts in our body, but this flexibility comes at a cost. The region is hugely vulnerable as it carries all the lifelines between the head and the body. In particular, these vertebrae are in close proximity to the two vertebral arteries, which are threaded through pairs of holes on either side of each vertebra. This is illustrated in Figure 6.

Before they supply oxygen-rich blood to the brain, each artery takes a sharp twist because of the structure of the topmost vertebra. This kink in the arteries is perfectly natural and causes no problems, except when the neck is stretched and simultaneously turned in an extreme or sudden way. This can occur when chiropractors perform their hallmark high-velocity, low-amplitude thrust manipulation. Such an action can result in a

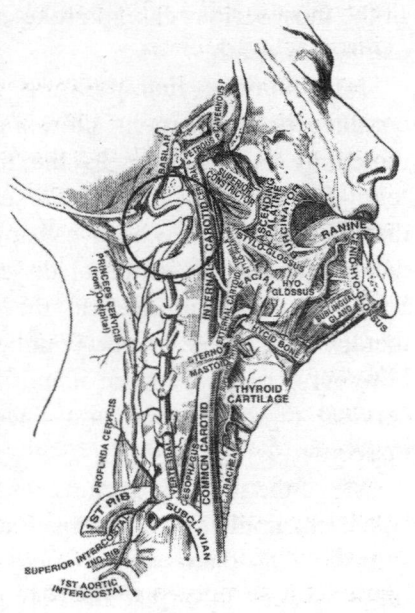

Figure 6
A circle shows how a vertebral artery makes a sharp kink at the final vertebra.

so-called vertebral dissection, which means that the inner lining of the artery becomes torn. Vertebral dissection can affect blood flow in four ways. First, a blood clot can form around the damaged area, gradually blocking that section of the artery. Second, the clot can eventually become dislodged, be carried into the brain and block a distant part of the artery. Third, blood can become trapped between the inner and outer layers of the artery, which causes swelling, thereby also reducing the blood flow. Fourth, the injury to the artery can cause it to go into spasm; this means it contracts and effectively prevents blood flow. In all four situations, vertebral dissection can ultimately cut off the blood supply to some parts of the brain, which in turn would lead to a stroke.

In the most serious cases, stroke can lead to permanent brain damage or death.

Unfortunately, manipulating the cervical region is a common practice among chiropractors, because it was promoted by B. J. Palmer as the most powerful chiropractic 'cure-all'. Chiropractors have been oblivious to the damage that they might have been causing, because there is often a delay between the vertebral dissection and the blockage of blood to the brain. Hence, the link between chiropractic therapy and strokes went unnoticed for decades. Recently, however, cases have been identified where manipulating the cervical region has certainly been the cause of vertebral dissection.

One of the most clear-cut cases highlighting the dangers of spinal manipulation concerns Laurie Mathiason, a twenty-year-old Canadian, who visited a chiropractor twenty-one times between July 1997 and February 1998 in order to relieve her low back pain. On each occasion the chiropractor manipulated her cervical vertebrae, but after her penultimate visit she complained of pain and stiffness in her neck. That evening she became clumsy, dropping ashtrays and plates at the restaurant where she worked, so she returned to the chiropractor the next day.

As the chiropractor manipulated her neck once again, Laurie began to cry, her eyes started to roll, she foamed at the mouth, her body began to convulse and she turned blue. She was rushed to hospital, slipped into a coma and died three days later.

The sudden and unusual nature of Laurie's death led to an inquest, with the aims of establishing the circumstances of her death and preventing similar tragedies in the future. After four days of evidence, it seemed that the penultimate chiropractic treatment had almost certainly damaged Laurie's vertebral artery, which caused a blood clot in one of the arteries supplying her brain and led to the relatively mild effects she felt that evening. The final treatment then dislodged the blood clot, which subsequently travelled to Laurie's brain and killed her.

The coroner's jury attributed Laurie's death to 'traumatic rupture of the left vertebral artery', and the coroner declared: 'So at this point, the public knows that Laurie died of a ruptured vertebral artery, which occurred in association with a chiropractic manipulation of the neck.' The jury also suggested a series of measures aimed at minimizing the risks to patients, which initially seemed to be well received by the senior figures in the chiropractic community. Unfortunately, this positive reaction from chiropractors rapidly evaporated, as they failed to implement the recommendations of the inquest and began to deny responsibility for Laurie's death.

Two days after the inquest ended, Paul Carey, president of the Canadian Chiropractic Protective Association, boldly stated on CBC Radio: 'The jury members did not make a direct relationship to the chiropractic adjustment.' Just a few weeks later, a press release from the Canadian Chiropractic Association claimed: 'The jury did not make a finding that chiropractic treatment was the cause of this tragedy.' These statements were followed by chiropractic newspapers, newsletters, leaflets and advertisements that seemed to contradict the findings of the inquest, which only added to the grief felt by the Mathiason family. Laurie's mother, Sharon Mathiason, was openly critical when these pronouncements appeared:

> I say that what chiropractors are doing is waging a co-ordinated, intentional campaign of fraud and deceit on the Canadian public. This does not allow anyone who is contemplating going to a chiropractor to have a full and accurate truth about Laurie's death. People are not being properly informed of the risk of chiropractic.

The case of Laurie Mathiason is certainly not unique. Indeed, in Canada alone there have been several other young women, such as Dora Labonte, Lana Dale Lewis and Pierrette Parisien, who have died soon after receiving chiropractic manipulation of the neck. These high-profile cases have made headlines and generated major discussions about the safety of chiropractic manipulation, but the key question is whether these tragedies are freak accidents, perhaps involving patients who were already vulnerable to strokes, or whether they are the tip of an iceberg, hinting at a widespread risk for all patients.

There have been several attempts to assess the level of risk associated with chiropractic neck manipulation, and the one that is most often quoted by chiropractors is a study entitled *The Appropriateness of Manipulation and Mobilization of the Cervical Spine*. Conducted in 1996, it estimated that the number of strokes, cord compressions, fractures and large blood clots was 1.46 per million neck manipulations. This is a remarkably small risk, essentially one in a million, but it is misleadingly low for two reasons. First, experts suspect that the vast majority of incidents go unreported and unrecognized, so most incidents could not have been included in the study. Second, a patient may receive more than ten treatments for a particular condition, thus increasing the risk by a factor of more than ten.

Other surveys have suggested higher risks, and perhaps the most significant study was published by Canadian researchers in 2001, concluding that the risk of artery damage is one incident per 100,000 individuals receiving chiropractic neck manipulation. They compared patients who had suffered damage to their vertebral arteries with control groups with no history of stroke. The results showed that patients under forty-five years of age who had suffered torn arteries were five times more likely to have visited a chiropractor in the week prior to the damage being recognized than healthy individuals of a similar age. This implies that chiropractic treatment can increase the risk of damaged arteries by a factor of five.

One of us, Professor Ernst, has repeatedly reviewed the literature on the risks of spinal manipulation. To date, about 700 cases of serious complications have been documented in these publications. This should be a major concern for health officials, particularly as under-reporting will mean that the actual number of cases is much higher. Indeed, if spinal manipulation were a drug with such serious and widespread adverse effects and so little demonstrable benefit, then it would almost certainly have been taken off the market by now.

The risk of torn arteries caused by chiropractors, and the dire consequences of such damage, raises three serious criticisms of the chiropractic profession. First, it is surprising that the exact risk associated with spinal manipulation is so poorly understood. Chiropractors seem to have no system for recording and monitoring the damage that they may inadvertently cause, and therefore they seem to be avoiding any attempt to assess the safety of their profession. This problem was highlighted in 2001, when a team of researchers, including Edzard Ernst, asked members of the Association of British Neurologists to report cases of neurological complications referred to them that had occurred within twenty-four hours of neck manipulation. They identified thirty-five cases, which included nine strokes, over the course of one year. Ernst and his colleagues were shocked to find that none of these cases had hitherto attracted any attention, inasmuch as they had not been reported in the medical literature or anywhere else.

The approach of the chiropractic profession stands in stark contrast to the conventional medical establishment, which rigorously assesses the safety of drugs before they are made publicly available. Even when a drug is available for prescription, doctors are encouraged to continue to monitor and report any adverse incidents in order to identify any rare side-effects. In Britain, this programme of ongoing vigilance is called the Yellow Card Scheme and it is administered by the Medical Healthcare Products Regulatory Agency (MHRA). This and other methods are the reasons why we learn about hitherto unknown risks and why we can, if risks emerge, withdraw a drug. Nothing remotely similar exists in the world of chiropractic.

The second criticism of chiropractors is that they often fail to warn patients of the potential risks of their treatment. A study published by chiropractors in 2005 assessed the consent policy among 150 randomly selected chiropractors in the United Kingdom, and it revealed that only 23 per cent always discussed serious risks with their patients before treatment. This contravenes the requirements of the Department of Health in the UK: 'Before you examine, treat or care for competent adult patients you must obtain their consent . . . Patients need sufficient information before they can decide whether to give their consent: for example information about the benefits and risks of the proposed treatment, and alternative treatments.' It also falls short of the Code of Practice of the British General Chiropractic Council: 'Before instituting any examination or treatment, a chiropractor shall ensure that informed consent to such treatment or examination has been given. Failure to obtain informed consent may lead to criminal or civil proceedings.'

The third criticism is that chiropractors continue to offer treatments for non-musculoskeletal conditions, even though spinal manipulation can have no impact on these conditions. This is an issue of concern that has already been raised, but it becomes even more problematic if we bear in mind the two previous criticisms. Not only is spinal manipulation useless in terms of treating, for example, asthma, but it also carries a potentially deadly risk and patients are not always informed of this.

Earlier in this chapter we offered some words of caution to patients who might be considering seeing a chiropractor, but we would like to add to this advice in light of the serious risks that we have now outlined. For example, we have stated that chiropractic therapy can sometimes help with back problems, and therefore short-term treatment from a mixer chiropractor can be helpful, but we have also stated that chiropractors tend to be no better than conventional physiotherapists at treating such conditions. Hence, because physiotherapeutic exercise is a much safer treatment than chiropractic manipulation, we would strongly recommend the former rather than the latter as the first choice.

Another treatment option, which we would also recommend ahead of chiropractic therapy, is *osteopathy*. The origins of osteopathy are similar to those of chiropractic therapy, inasmuch as both emerged in North America at the end of the nineteenth century as the result of discoveries made by charismatic mavericks. In the case of osteopathy, the founder was Andrew Taylor Still. He believed that manipulating bones in general, not just the spine, improved blood flow and enhanced the nervous system. Moreover, he argued that manipulating bones could enable the body to heal every possible illness!

Although the traditional ambitions of both chiropractic therapy and osteopathy – manipulating the spine or bones in general can cure everything – are equally bizarre and misguided, we would advise the latter rather than the former for several reasons. First, osteopaths have largely shed the more bizarre beliefs and claims from their early days and are today more solidly grounded in science. Second, they usually employ gentler techniques which cause fewer adverse side-effects. Third, they use fewer X-rays and are less likely to employ untested diagnostic methods. Fourth, they generally do focus on conditions relating to the spine and the musculoskeletal system, leaving other diseases to other specialists. Importantly, however, *cranial osteopathy* is not a treatment that we would recommend as there is no evidence at all that it works. Both osteopathy and cranial osteopathy are explained in more detail in the appendix.

If you do still decide to visit a chiropractor, despite all our concerns and warnings, then we would very strongly recommend that you start your consultation by confirming that your chiropractor will not be manipulating your neck. Even if your problem relates to your lower back, it is still worth stressing that your neck should not be touched, as many chiropractors will manipulate the neck without informed consent in order to address a whole range of conditions. Indeed, Laurie Mathiason, who died in 1998, was having her neck manipulated even though she was concerned about a pain in her lower back.

Finally, before ending this section on the risks of chiropractic therapy, it is important to stress that all the concerns that have been raised also apply to the treatment of children. Many parents feel that they are acting in their children's best interests by taking them to see a chiropractor, but they should be aware that they are exposing them to the hazards of X-rays, temporary adverse reactions, spinal injuries and even stroke. In fact, the dangers to children are particularly worrying, because children's bones continue growing right to the end of their teenage years, so a chiropractor would be manipulating an immature spine.

As with adults, many chiropractors are willing to treat children for wholly inappropriate conditions, such as asthma, bedwetting, clumsiness, ear infections, gastric problems, hyperactivity, immune-system problems, learning disorders and respiratory problems. Chiropractors will claim to be able to treat such conditions, but we know that the evidence does not suggest that spinal manipulation can offer any benefit.

Prompted by these sorts of unfounded claims, the journalists Paul Benedetti and Wayne MacPhail have investigated the issues surrounding children and chiropractic therapy, and they included their findings in *Spin Doctors: The Chiropractic Industry Under Examination*. They focused their attention on their native Canada, where virtually all chiropractors deal with children, and where a significant fraction of parents seek out chiropractic treatment for their offspring. Indeed, according to a survey conducted in 2004 in Toronto, 31 per cent of children had been treated by a chiropractor.

To find out what happens to children who visit a chiropractor, Benedetti and MacPhail arranged for a colleague to accompany an eleven-year-old-girl, known as Judy, to visit five chiropractors in the Toronto area in 2001. Judy was also examined by Dr John Wedge, an experienced paediatric orthopaedic surgeon at Toronto's Hospital for Sick Children, who confirmed that she was 'a perfectly healthy girl'. The goal was to find out if the chiropractors would agree with Dr Wedge's conclusion.

The chiropractors were told that Judy was generally in good health, but that she suffered from a few earaches, some mild headaches and allergies, and that there was concern about the possibility of asthma. One chiropractor examined Judy and agreed that she was fine and recommended no spinal manipulation, but the other four found a whole range of problems. Moreover, different chiropractors found different problems and gave conflicting diagnoses.

According to Benedetti and MacPhail: 'The other four put her through a barrage of tests and found imbalances, partially locked vertebrae, asymmetry, uneven weight distribution, and a spinal column riddled from top to bottom with subluxations. They found subluxations in the upper, middle and lower spine, but not necessarily in the same vertebrae.' The chiropractors claimed that the subluxations that they had identified could lead to learning problems, digestive ailments and reproductive difficulties, and one of them claimed to detect signs of early osteoarthritis. Not surprisingly, these chiropractors recommended spinal adjustment, with one of them proposing immediate care in the form of six visits a week for two weeks, then three times a week for six weeks, followed by twice weekly visits until the problems had improved.

The journalists transcribed the comments of the chiropractors. One suspected that Judy's problems could be traced back to her birth: 'The surgeon or somebody grabbing her head, twisting it one way or the other. A lot of stress was created. I would estimate probably 85 to 95 per cent of all the problems I see in adults start from the process of delivery if you can believe it.'

Another chiropractor took a thermal scan of Judy's back and phoned her uncle later to explain the results: 'Okay. I can tell you that her scan is horrible. Her thermography scan is terrible. From the top of her neck all the way down to the upper part of her low back is showing nerve interference. That's a huge area in someone her age. Now, I haven't seen her X-rays obviously because I don't think you've had them done yet, is that correct?' Although the uncle explained that Judy's mother did not want her daughter to be X-rayed, the chiropractor tried to change her decision: 'Okay, her mum would need to speak to me then. But I really think it's necessary, especially seeing what I'm seeing on the scan and what I saw in the exam.'

Encouraging a child to have unnecessary X-rays, raising unjustified concerns about serious conditions and offering to manipulate immature bones all reflect badly on the chiropractic profession. Benedetti and MacPhail, however, uncovered an even more disturbing practice, namely the chiropractic manipulation of babies. One of their colleagues pretended to be the mother of a two-year-old baby prone to ear infections and called fifty chiropractors selected at random from the Toronto phone directory. Her survey revealed that 72 per cent offered to treat her fictional child, even though there is no evidence that chiropractors can help with ear infections.

The risks of alternative medicine

Most people view alternative medicine as a safe option. On the other hand, conventional medicine is often criticized because of the side-effects of pharmaceutical drugs or the risks associated with surgery. But is alternative medicine really safer than conventional medicine?

We have already seen that chiropractic therapy carries a range of risks, from the minor hazard associated with X-rays to the strokes that can be caused by manipulation of the upper spine. In short, chiropractic therapy is certainly more dangerous than conventional physiotherapeutic exercise. But, what about the other alternative therapies?

In the previous chapters we deliberately omitted the issue of safety in the context of acupuncture and homeopathy, as we wanted to focus initially on whether or not these therapies were effective. However, now that we have raised the issue of safety, we will discuss the risks associated with these treatments in the remainder of this chapter. For both these therapies, the primary issues are an assessment of the risk, determining whether or not the benefit outweighs the risk, and comparing the risk/benefit ratio with the risk/benefit associated with conventional medicine.

In the case of acupuncture, studies have shown that treatments can result in minor pain, bleeding or bruising, but these adverse reactions are only minor: they occur in roughly 10 per cent of patients and are transient. Slightly more serious side-effects include fainting, dizziness, nausea and vomiting, but again these events are rare and are usually associated with anxious patients who may have a fear of needles. Although most patients may accept such risks as an unsurprising consequence of being pierced with needles, there are two very serious adverse effects that patients should consider in advance of visiting an acupuncturist.

The first of these is infection. This is a concern because there have been several documented cases of patients contracting diseases such as hepatitis. For example, the journal *Hepatology* documented how 35 out of 366 patients contracted hepatitis B from an acupuncture clinic in Rhode Island. A detailed study of this outbreak showed that patients who received fewer than 150 needle insertions during the course of their treatment ran a 9 per cent risk of contracting the disease, whilst those who received more than 450 needles ran a 33 per cent risk. The infection is caused by re-using needles that have not been properly sterilized, and part of the problem may be due to the Chinese tradition of storing needles in alcohol solutions, which is not sufficient to protect against the hepatitis virus.

The other serious risk to patients is the danger that needles might puncture and damage a major nerve or organ. For example, needling at the base of the skull can lead to brain damage, deep needling in the lower back can damage a kidney, and there are over sixty reported cases of punctured lungs, known as *pneumothorax*. Most worrying of all, there is a report of an acupuncturist inserting a needle in the chest of an Austrian female patient and penetrating her heart. Normally needling at this point is entirely safe because the sternum protects the heart, but one in twenty people have a hole in that bone. This abnormality cannot be felt or seen because it is covered by very strong ligaments, but an acupuncture needle will go right through these structures. In the case of the Austrian patient, it pierced the heart and killed her.

Although acupuncture carries some common and serious

risks, it is important to stress that the common risks are not at all serious and the serious risks are not at all common. The sixty cases of pneumothorax reported in recent decades have to be appreciated in the context of the millions of acupuncture treatments given each year. Moreover, the serious risks can be minimized by visiting a medically trained acupuncturist who uses disposable needles.

On the other hand, it should be remembered that the evidence for the efficacy of acupuncture ranges from zero for a whole range of conditions to borderline for some types of pain relief and nausea. Hence, it would only be worth considering acupuncture for treating pain relief and nausea, and only then if you feel on balance that the supposed benefits are large enough to outweigh the small risks.

In Chapter 3 we discussed the efficacy (or rather the lack of efficacy) of homeopathy. The conclusion was that the benefits of homeopathy were purely placebo, which is not surprising bearing in mind that the final medicine often contains no active ingredient because of the extreme levels of dilution. One might assume, therefore, that homeopathy would at least be safe. Surely, if homeopathic remedies are devoid of any active ingredient then they must be harmless?

Unfortunately, homeopathy can have surprising and dangerous side-effects. These have nothing to do directly with any particular homeopathic remedy, but rather they are an indirect result of what happens when homeopaths replace doctors as sources of medical advice.

For example, many homeopaths have a negative attitude towards immunization, so parents who are in regular contact with a homeopath may be less likely to immunize their child. To evaluate the extent of this problem, Edzard Ernst and Katja Schmidt at Exeter University conducted a revealing survey among UK homeopaths. Having obtained e-mail addresses from online directories, they sent an e-mail to 168 homeopaths in which they effectively pretended to be a mother asking for advice about whether or not to vaccinate her one-year-old child against measles, mumps and rubella (MMR). This was in 2002 when the controversy over MMR was subsiding and the scientific evidence was clearly in favour of vaccination. Although 104 homeopaths replied, the ethics committee overseeing the research survey required that these homeopaths be informed of the real purpose behind the e-mail and be given the opportunity to withdraw their replies if they were unwilling to be involved in the survey. Sure enough, twenty-seven homeopaths took advantage of this opportunity. Of the remaining seventy-seven respondents, only two (or 3 per cent) advised the mother to immunize. Of course, the responses of the twenty-seven homeopaths who withdrew from the survey have never been made public or evaluated, but it seems reasonable to assume that their general attitude on average would have been even more negative. It is clear that the overwhelming majority of homeopaths will not encourage immunization.

This anti-immunization stance is not unique to homeopaths, but it is also common among other alternative therapists. At the same time as surveying homeopaths, Ernst and Schmidt also emailed chiropractors with the same request for immunization advice. There were twenty-two responses, but six chiropractors withdrew when they realized their replies were part of an academic survey. Out of the remaining sixteen responses, only four (25 per cent) chiropractors advised immunization. Again, we can assume that those who withdrew from the survey would have had an even more negative attitude. And, again, it is clear that the vast majority of chiropractors will not encourage immunization.

This negative response from chiropractors was in keeping with the openly hostile attitude that is expressed in much of their literature. Senior chiropractors have written statements such as 'smallpox vaccination was stopped in the US and UK because it was realized that the vaccinated suffered the worst effects of the disease' and 'the dangers of vaccination to the young child are profound . . . in some cases, the vaccine acts non-specifically to increase a child's pre-existing chronic disease tendency'. Both these statements are misleading and damaging. The truth is that immunization is arguably the single most important discovery in the history of medicine. Indeed, there is probably a substantial proportion of readers, possibly including you, who would not be alive today if it were not for the immunizations that we all received when we were children.

Thankfully these diseases are now rare in the developed world, but this means that it is easy to forget their potentially devastating impact – we no longer appreciate why we used to fear them. If, however, we look beyond the developed world then we can be reminded of the dangers of childhood diseases and the value of vaccination. For example, the Measles Initiative was launched in 2001 to vaccinate children and reduce deaths from measles around the world. In its first five years the programme has already reduced the annual number of deaths from measles in Africa by 91 per cent, from over 400,000 to 36,000.

The widespread anti-immunization stance among alternative therapists is just one of the ways in which they offer harmful advice to patients. Another example is that alternative therapists sometimes meddle with a patient's conventional drug-treatment programme, even though they are not qualified to advise about a patient's prescription. A 2004 survey of UK-based acupuncturists showed that 3 per cent of patients received advice about their prescriptions, some of whom suffered adverse consequences.

Perhaps the greatest danger in the way that alternative therapists behave is simply the promotion of their own treatments when patients should be in the care of a conventional doctor. There are numerous reports of patients with serious conditions (e.g. diabetes, cancer, AIDS) suffering harm after following irresponsible advice from alternative practitioners instead of following the advice of a doctor.

This danger is amplified by a particularly strange facet of many alternative therapies, a phenomenon known as the *healing crisis*. This means that an expected part of the healing process is that the therapy might cause symptoms to deteriorate before they improve – this is supposedly due to the body fighting back or toxins being expelled. In one case, a patient being treated for pancreatitis (a life-threatening condition) was given a homeopathic remedy with a label advising that abdominal pain was part of the healing crisis, otherwise known as a *homeopathic aggravation*. So, just when the pancreatitis might be worsening and the patient ought to be seeking urgent medical attention, the homeopathic advice would be that the patient should relax because everything is progressing as expected.

In 2006, Simon Singh, one of the authors of this book, attempted to highlight the extent to which homeopaths give bad advice by finding out what they would offer to a young traveller seeking protection against malaria. Working with Alice Tuff and the charity Sense About Science, Singh developed a storyline in which Tuff would be making a ten-week overland trip through West Africa, where there is a high prevalence of the most dangerous strain of malaria, which can result in death within three days. The idea was that Tuff, a young graduate, would explain to homeopaths that she had previously suffered side-effects from conventional malaria tablets and wondered if there was a homeopathic alternative.

Before approaching homeopaths, however, Tuff visited a conventional travel clinic with exactly the same storyline, which resulted in a lengthy consultation. The health expert explained that side-effects were not unusual for malaria tablets, but that there was a range of options, so a different type of tablet might be advisable. These tablets could then be taken in the week prior to departure to check that they had no unpleasant side-effects. At the same time, the health expert asked detailed questions about Tuff's medical history and offered extensive advice, such as how to prevent insect bites.

Tuff found a variety of homeopaths by searching on the internet, just as any young student might do. She then visited or phoned ten of them, mainly based in and around London. Some of these homeopaths ran their own clinics, some were based at homeopathic pharmacies and one was based in a major mainstream pharmacy. In each case, Tuff secretly recorded the conversations in order to document the consultation.

The results were shocking. Seven out of the ten homeopaths failed to ask about the patient's medical background and also failed to offer any general advice about bite prevention. Worse still, ten out of ten homeopaths were willing to advise homeopathic protection against malaria instead of conventional treatment, which would have put our pretend traveller's life at risk.

The homeopathic remedies differed between homeopaths. Some offered Malaria nosode (based on rotting vegetation), while others recommended China officinalis (based on quinine) or Natrum muriaticum (based on salt). In every case, the remedies were so diluted than none of them contained any active ingredient and all were equally useless.

The homeopaths offered anecdotes to show that homeopathy is effective. According to one practitioner, 'Once somebody told me she went to Africa to work and she said the people who took malaria tablets got malaria, although it was probably a different subversive type not the full blown, but the people who took homeopathics didn't. They didn't get ill at all.' She also advised that homeopathy could protect against yellow fever, dysentery and typhoid. Another homeopath tried to explain the mechanism behind the remedies: 'The remedies should lower your susceptibility; because what they do is they make it so your energy – your living energy – doesn't have a kind of malaria-shaped hole in it. The malarial mosquitoes won't come along and fill that in. The remedies sort it out.'

A few days later the BBC programme *Newsnight* conducted undercover filming at some of the same homeopathic clinics and found exactly the same empty remedies being offered to protect against malaria. It was in the run-up to the summer holiday season, so this became part of a campaign to warn travellers against the very real dangers of relying on homeopathy to protect against tropical diseases. One case reported in the *British Medical Journal* described how a woman had relied on homeopathy during a trip to Togo in West Africa, which resulted in a serious bout of malaria. This meant she had to endure two months of intensive care for multiple organ system failure.

The main point of the investigation into the homeopathic treatment of malaria was to demonstrate without doubt that even the most benign alternative medicine can become dangerous if the therapist who administers it advises a patient not to follow an effective conventional medical treatment.

It is likely that some of the alternative therapists who sell useless remedies for dangerous conditions are fully aware of what they are doing and are happy to profit from it. Before ending this chapter, however, it is important to stress that the majority of alternative therapists are acting with the best of intentions. These misguided therapists are simply deluding themselves, as well as their patients.

One of the most poignant examples of a well-intentioned homeopath is the case of an English homeopath working in Devon, whose identity cannot be revealed. In 2003, she noticed a brown spot on her own arm, which was growing in size and changing in colour. At the time, she was in regular contact with doctors as she was taking part in a study organized by Professor Ernst, which was designed to see if homeopaths could treat asthma. Rather than discussing her lesion with the doctors, however, she decided to treat it herself using her own homeopathic remedies.

She had such faith in her remedies that she treated the spot for several months and continued to keep it a secret from the doctors. Unfortunately, the spot turned out to be a malignant melanoma. As each month passed, the chance for early treatment of this aggressive form of cancer steadily disappeared. While she was still in the middle of treating asthma patients, the homeopath died. Had she sought conventional treatment at an early stage, then there might have been a 90 per cent chance that she would have survived for five or more years. By relying on homeopathy, she had condemned herself to an inevitably early death.

5 The Truth About Herbal Medicine

'The art of healing comes from nature and not from the physician. Therefore, the physician must start from nature with an open mind.'

Paracelsus (1493–1541)

Herbal Medicine

The use of plants and plant extracts in the treatment and prevention of a whole range of diseases.
Herbal medicine is one of the oldest and most
widespread forms of treatment. Based on local
plants and traditions, it continues to play a major
role in healthcare in Asia and Africa. In recent
decades, herbal medicine, sometimes known as
phytotherapy, has become one of the fastest-
growing forms of treatment in the rest of the
world.

THE FIRST CASE OF ALTERNATIVE MEDICINE DISCUSSED IN THIS book concerned Ötzi, the 5,000-year-old mountain hiker whose frozen body was found in Austria in 1991 with various tattoo marks. These tattoos were located at points that are still familiar to modern acupuncturists, so it seems possible or even likely that Ötzi had been receiving a treatment akin to acupuncture. There is additional evidence, however, that Ötzi was also receiving another form of alternative medicine, namely herbal medicine.

Archaeologists studying Ötzi's body found two walnut-sized lumps threaded together on a leather thong. The lumps were identified as the fruit of the birch fungus (*Piptoporus betulinus*), which contains polyporenic acid, which acts as an antibiotic. This discovery became particularly interesting when scientists discovered that Ötzi's colon was infected with the eggs of a parasitic whipworm known as *Trichuris trichiura*, which can be killed by polyporenic acid. Writing in the *Lancet*, an anthropologist named Dr Luigi Capasso concluded: 'The discovery of the fungus suggests that the Iceman was aware of his intestinal parasites and fought them with measured doses of *Piptoporus betulinus*.'

From Ötzi's fungal medication and similar archaeological evidence, we know that mankind's most ancient system of medicine was based on plants. Of course, our ancestors would have had no idea that *Piptoporus betulinus* contained polyporenic acid and that this killed the eggs of *Trichuris trichiura*, but they knew enough to realize that consuming birch fungus somehow alleviated certain types of stomach pains, and similarly they worked out that other plants somehow cured other conditions.

Societies around the world used trial and error to develop their own bodies of medical knowledge based on local plants, with the tribal healer acting as the expert database and provider of medicines. Each generation of sangomas and shamans gradually accumulated further information about the natural remedies that grew around them, so that herbal medicine became an increasingly powerful system of healthcare. Then, in the eighteenth century, herbal medicine suddenly entered a new era when it started to be investigated by scientists who sought to improve on nature's medicine cabinet.

In 1775 a British physician called William Withering joined the staff at Birmingham General Hospital and shortly thereafter he became a regular attendee at the Lunar Society. The Lunatics were a group of eminent men who met once a month on the Monday closest to the full moon – this allowed them to discuss science late into the night and still have some illumination on their journey home. Withering's career in medicine combined with his interest in science resulted in a major investigation into the medical benefits of the foxglove plant, also called *digitalis*. It had long been known that digitalis could be used to treat dropsy, a swelling associated with congestive heart failure, but Withering spent nine years meticulously documenting its impact on a total of 156 patients. In his experiments, he varied how the digitalis was prepared and also altered the dosages in order to learn how to maximize the herb's benefits and minimize its side-effects. For example, he learned that the dried powdered leaf was five times more effective than a fresh foxglove leaf; that boiling the leaf weakened its impact on the patient; and that excessive use of the plant would lead to nausea, vomiting, diarrhoea and a tendency to see the world with a yellow-green tinge.

He published his research in 1785 in a book entitled *An Account of the Foxglove and Some of its Medical Uses*. His report highlighted his rigorous and impartial approach to analysing digitalis:

> It would have been an easy task to have given select cases, whose successful treatment would have spoken strongly in favour of the medicine, and perhaps been flattering to my own reputation. But Truth and Science would condemn the procedure. I have therefore mentioned every case . . . proper or improper, successful or otherwise.

Withering's research marks a turning point in the history of herbal medicine, from its haphazard ancient roots towards a more systematic and scientific attitude. One by one, traditional herbs were submitted to scrutiny. A good illustration of this new rational approach is the way that scientists harnessed the potential of cinchona tree bark, which had long been used by the Peruvian Indians to treat malaria. Jesuit priests learned of its curative powers in the 1620s, and within a couple of decades the so-called Jesuit's Bark was highly valued in large parts of Europe. Indeed, the seventeenth-century Italian doctor Sebastiano Bado considered cinchona bark to be a more valuable treasure than all the gold that had been brought back from South America.

Herbalists prepared the cinchona bark for medical use by simply drying it and then grinding it into a fine powder. It was this powder that inspired Samuel Hahnemann to invent homeopathy, as discussed in Chapter 3. Scientists, however, took the herbal remedy in quite a different direction and ultimately maximized its potential. Speculating that it was only one component of the bark that was medically active, they attempted to isolate that component and then deliver it in a more concentrated and potent manner. It took until 1820 before two French chemists, Pierre-Joseph Pelletier and Joseph-Bienaimé Caventou, isolated a compound that they called *quinine*, based on the Inca word for the cinchona tree. Thereafter, scientists could properly study in detail the effects of this anti-malarial substance and optimize how it could be used to save lives.

Just a few years after quinine was isolated from cinchona bark, scientists focused their attention on willow bark, which had been used to reduce pain and fevers for thousands of years. Once again, they successfully identified the active ingredient, this time naming it *salicin*, based on *salix*, the Latin word for willow. In this case, however, chemists took nature's drug and attempted to modify and improve it, driven by the knowledge that salicin was toxic. Taken in either its pure form or in willow bark, salicin was known to cause particularly harmful gastric problems, but chemists realized that they could largely remove this side-effect by transforming salicin into another closely related molecule known as acetylsalicylic acid. The Bayer Company in Germany started marketing this new wonder drug under the name of aspirin in 1899, and kicked off its promotional campaign by writing to 30,000 doctors across Europe in the first mass mailing in pharmaceutical history. Aspirin was an immediate success and there were numerous celebrity endorsements – Franz Kafka said to his fiancée that it eased the unbearable pain of being.

Thanks to the scientific approach, aspirin has gone from strength to strength. It is now the cheapest and biggest-selling drug in the world, and it has become far more than the painkiller it was first believed to be. Clinical trials have shown that it can reduce the risk of heart attack, stroke and many types of cancer. On the negative side, scientific investigations have also revealed that aspirin can lead to stomach bleeding in 3 out of every 1,000 people and can increase the risk of asthma attacks. Moreover, aspirin is not recommended for children under twelve years of age.

It is already becoming quite clear that this chapter on herbal medicine will be very different from the previous chapters on acupuncture, homeopathy and chiropractic manipulation. These other therapies have struggled to be accepted by mainstream medicine, partly because their underlying philosophies conflict with our scientific understanding of anatomy, physiology and pathology. Why should needling non-existent meridians improve hearing? Why should ultra-dilute homeopathic solutions devoid of any active ingredient treat hay fever? Why should manipulating the spine alleviate asthma? By contrast, plants contain a complex cocktail of pharmacologically active chemicals, so it is not surprising that some of them can impact on our wellbeing. Consequently, herbal medicine has been embraced by science to a far greater extent than the other treatments mentioned above.

Indeed, there is general agreement that much of modern pharmacology has evolved out of the herbal tradition. According to the neuroscientist Patrick Wall, 95 per cent of the painkillers used by today's doctors are based on either opium or aspirin, and the range of modern drugs based on plants includes the anti-cancer agent taxol (from the Pacific yew tree) and the anti-malarial drug *artemisinin* (from the artemisia shrub). Some nature-based remedies have had very humble origins, such as *penicillin*, which was discovered when a speck of *penicillium* mould floated into a laboratory in Paddington, London. Other remedies had to be tracked down in more exotic locations, such as Madagascar, home to a species of the periwinkle that has yielded dozens of interesting chemicals, including the drugs *vincristine* and *vinblastine* used in chemotherapy.

Despite all these examples, which demonstrate that numerous herbs have become part of mainstream medicine, it is important to stress that much of herbal medicine is still considered alternative. In fact, it is easy to make a division between alternative herbal medicine and what might be called scientific herbal medicine. The difference between the two categories becomes clear if we revisit the objectives of the scientists who examined plant-based remedies in the nineteenth and twentieth centuries.

The scientists wanted to identify the active ingredient of each plant and isolate it. Then they sought to synthesize it industrially, in order to mass produce it at a low cost. They even endeavoured to improve on nature by manipulating the molecules of the original ingredient. Crucially, the scientists attempted to evaluate the impact of their treatments on patients to find out which herbal extracts were safe and effective, and which were dangerous or ineffective. The treatments that emerged from this scientific approach to herbal medicine are so utterly mainstream that they are no longer labelled herbal medicines, but rather they are simply incorporated within the realm of modern pharmacology. It is certainly appropriate that the word *drug* comes from the Swedish word *druug*, meaning 'dried plant'.

On the other hand, alternative herbal medicine generally places an emphasis on using the whole plant or a whole part of the plant, because its underlying philosophy is that these plants have been designed to cure us. Traditional herbalists believe that Mother Nature has engineered the complex mix of substances within a plant so they work in harmony, which means that the plant produces an effect that is greater than the sum of its parts. Herbalists call this *synergy*.

In short, alternative herbal therapists continue to believe that Mother Nature knows best and that the whole plant provides the ideal medicine, whereas scientists believe that nature is just a starting point and that the most potent medicines are derived from identifying (and sometimes manipulating) key components of a plant.

We know that scientific plant-based pharmaceuticals are effective, but the key issue in the context of this book is whether or not alternative complete-plant herbal medicines actually work. Most have not been submitted to the same level of scrutiny as conventional drugs, but there are numerous studies that do shed light on particular herbal medicines. In the next section we have done our best to collate the evidence in order to decide whether each herb is genuinely effective – e.g. does echinacea cure a cold, and can evening primrose oil alleviate eczema?

We will also address an even more important issue, namely safety. As well as knowing which herbal medicines work, patients also need to know which are dangerous and perhaps even lethal.

The herbal pharmacy

Over the last two decades, there has been a plethora of newspaper articles championing the benefits of herbal remedies derived from St John's wort, a plant that can supposedly act as an antidepressant. Indeed, sales of St John's wort rocketed in the 1990s, so much so that its consumption increased more quickly than any other popular herbal medicine. But is this sales boom justified? Can St John's wort really help patients with depression?

St John's wort (*Hypericum perforatum*), which originates from Europe, would have been recognized by the earliest farmers as a poisonous plant, as it could harm grazing livestock, leading to problems such as spontaneous abortion and even death. Perhaps its toxicity led to the practice of hanging St John's wort in houses to ward off evil spirits. In time, the tradition evolved so that people would hang the plant on St John's day, 24 June, soon after its yellow flowers have emerged. The association with the saint's day is how the plant

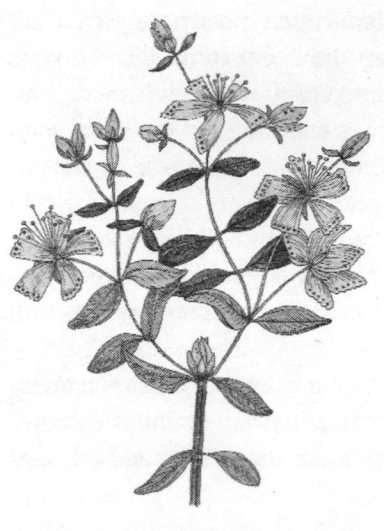

St John's wort

came to be called St John's wort, with the appendage *wort* merely being the Old English term for plant.

The notion that St John's wort could poison evil spirits in the outside world probably encouraged ancient healers to believe that the plant could also poison the evil spirits within us, which they believed were responsible for disease. We know that healers were using St John's wort to treat sciatica, arthritis, menstrual cramping, diarrhoea and many other conditions over 2,000 years ago, but it was not until the sixteenth century that the physician Paracelsus provided the first documented evidence indicating that the plant was being used to treat mental conditions, otherwise known as *phantasmata*. In the following century, an Italian doctor called Angelo Sala also described how this plant could be used to treat depression, anxiety and madness, and moreover he pointed out, 'St John's wort cures these disorders as quick as lightning.'

St John's wort continued to be used to treat depression right up to the start of the twentieth century, but it and other herbal remedies gradually fell out of favour as European and American doctors preferred to rely on new drugs that were being developed. Medicine was entering the scientific age, so there was an inevitable tendency to reject ancient natural remedies in favour of new pharmaceuticals. Nevertheless, the tradition of herbal medicine survived in pockets around

Europe and America, and there was always a steady stream of anecdotal evidence suggesting that St John's wort was effective for the treatment of depression. But did these stories of recovery indicate that St John's wort was genuinely effective, or could they be explained as the result of a powerful placebo effect?

The best way to determine the efficacy of St John's wort was to submit it to scientific testing, and from 1979 onwards there was a whole series of trials. Most of them originated in Germany, where herbal medicine had retained a core of sympathetic support among some doctors and patients. As often seems to be the case in alternative medicine, each individual trial was unable to give a definitive conclusion about the herb's efficacy, but over and over again there were tantalizing indications that St John's wort was more than a mere placebo. The next step was to perform a meta-analysis, whereby all the data from all the trials would be carefully brought together in order to get a firmer grasp of the true value of the plant.

The first meta-analysis of St John's wort was conducted in 1996 and included the results from twenty-three studies. Referring to St John's wort by its Latin name of *hypericum*, it concluded: 'There is evidence that extracts of hypericum are more effective than placebo for the treatment of mild to moderately severe depressive disorders.' In 1997, the American current affairs TV programme *20/20* said that St John's wort was 'a truly remarkable startling medical breakthrough – one that could affect millions of people who suffer from mild depression.' Thanks to this sort of publicity, sales of St John's wort in America increased by a factor of 30 in just three years.

The conclusion of the 1996 meta-analysis was reinforced in 2005 by the Cochrane Collaboration. It conducted a systematic review entitled *St John's wort for depression*, which covered all the thirty-seven trials that had emerged by then. In the context of treating mild or moderate depression, Cochrane stated, 'hypericum and standard antidepressants have similar beneficial effects'. However, the authors of the review did highlight the limitations of St John's wort: 'for major depression, several recent placebo-controlled trials suggest that the tested hypericum extracts have minimal beneficial effects'.

Nevertheless, the overall conclusion for St John's wort is still positive, as it offers similar benefits to modern drugs in treating mild to moderate depression. It is therefore another tool that can be used to help patients who may not respond to existing conventional drugs. There have been attempts to isolate the key active ingredient in St John's wort, thought to be either *hyperforin* or *hypericin*. When these have been tested, however, it appears that they are not as effective as the plant itself. In this particular instance, the herbalist's view seems to be correct. In other words, it seems that the benefits of St John's wort are due to a combination of chemicals, each one working to enhance the effect of the others.

Because it is backed by research, St John's wort has grown to be one of the biggest sellers in an annual worldwide market for herbal remedies that is now worth roughly £10 billion. Today's pharmacies and health shops offer hundreds of herbal medicines, each one usually promoted as a treatment for several conditions. With so many remedies and conditions, it would be impractical within the limitations of this book to examine each herb in the same level of detail that we have provided for St John's wort, but we are able to offer a brief verdict on all the bestselling herbal remedies.

Table 1 lists each herbal medicine along with the main conditions that it is used to treat. In each case, the herb is given one of three ratings depending on the research evidence that has been accumulated so far in support of its effectiveness. The ratings are good, medium and poor.

For example, we have given devil's claw a 'good' evidence rating for treating musculoskeletal pain because several high-quality trials have indicated that it is effective and the evidence is uniform – i.e. there are no significant studies that suggest the opposite.

Feverfew is given a 'medium' evidence rating for preventing migraine because there have been mixed results from trials – mainly positive, but partly negative. And the positive trials have not been wholly convincing due to the quality of the trials, the number of patients involved and the small effect observed.

Lavender is given only a 'poor' evidence rating in the treatment of insomnia and anxiety, because it has undergone very few trials and the results have been contradictory. Interestingly, some very high-profile herbal medicines, such as chamomile and evening primrose, are also classified as being supported only by poor evidence. The reputations of these herbal remedies are probably a result of clever marketing coupled with the placebo effect experienced by the purchasers. In short, it is likely that you would be better off spending your money on effective conventional medicines rather than on herbal medicines with a poor evidence rating.

Table 1 provides a good starting point for appreciating the efficacy of herbal medicines, but four important points need to be made in order to put it into context. First, even though some of the herbal medicines in the table seem likely to be effective for some conditions, there are conventional pharmaceuticals that offer equal or greater benefit in almost all cases. The only important exception is in the treatment of the common cold, because conventional medicines are largely ineffective and echinacea extracts have shown some positive results in trials. Although echinacea may not prevent the onset of a cold, it may be worth taking it during a cold because it possibly reduces the length of the illness.

Table 1 – The efficacy of herbal medicines

Each herb is followed by the conditions it supposedly treats and a rating. The ratings reflect the amount and quality of evidence supporting the efficacy of each herb. Those herbs given a poor rating should be avoided, as there is no good reason to believe that they are effective. Even those herbs given moderate and strong ratings are not necessarily advisable for patients – the reasons for caution are explained in the next section of this chapter.

It is worth noting that for many diseases and conditions, including cancer, diabetes, multiple sclerosis, osteoporosis, asthma, hangover and hepatitis, there are no effective herbal remedies.

Aloe vera (*Aloe barbadensis*): herpes, psoriasis, wound healing, skin injuries. — *Poor*

Andrographis (*Andrographis paniculata*): common cold. — *Medium*

Artichoke (*Cynara scolymus*): high cholesterol, dyspepsia. — *Poor*

Bilberry (*Vaccinium myrtillus*): eye conditions, varicose veins, phlebitis, menstrual pain. — *Poor*

Black cohosh (*Actaea racemosa*): menopause, cold, menstrual and other gynaecological problems. — *Medium*

Chamomile (*Chamomilla recuita*): a 'cure all' – e.g. dyspepsia, irritable bowel syndrome, insomnia. — *Poor*

Cranberry (*Vaccinium macrocarpon*): prevention of infections in the urinary-tract. — *Medium*

Devil's claw (*Hapargophytum procumbens*): musculoskeletal pain. — *Good*

Echinacea (*E. angustifolia, pallida, or purpurea*): treatment and prevention of common cold. — *Good*

Evening primrose (*Oenothera biennis*): eczema, menopausal problems, PMS, asthma, psoriasis; a 'cure all'. — *Poor*

Feverfew (*Tanacetum parthenium*): migraine prevention. — *Medium*

Garlic (*Allium sativum*): high cholesterol. — *Good*

Ginger (*Zingiber officinalis*): nausea. — *Medium*

Ginkgo (*Ginkgo biloba*): dementia, poor circulation in the leg.   *Medium*

Ginseng, Asian (*Panax ginseng*): impotence, cancer, diabetes;   *Poor*
a 'cure all'.

Ginseng, Siberian (*Eleutherococcus senticosus*): enhancement   *Poor*
of performance, herpes.

Grape seed (*Vitis vinifera*): prevention of cancer and   *Medium*
cardiovascular disease.

Hawthorn (*Crataegus spp.*): congestive heart failure.   *Good*

Hops (*Humulus lupulus*): insomnia.   *Poor*

Horse chestnut (*Aesculus hippocastanum*): varicose veins.   *Good*

Kava (*Piper methysticum*): anxiety.   *Good*

Lavender (*Lavendula angustifolia*): insomnia, anxiety.   *Poor*

Ma huang (*Ephedra sinica*): weight loss.   *Good*

Milk thistle (*Silybum marianum*): hepatitis and liver disease   *Medium*
caused by alcohol.

Mistletoe (*Viscum album*): cancer.   *Poor*

Nettle (*Urtica dioica*): benign prostate hyperplasia.   *Medium*

Passion flower (*Passiflora incarnata*): insomnia, anxiety.   *Poor*

Peppermint (*Mentha x piperita*): irritable bowel syndrome, dyspepsia.   *Medium*

Red clover (*Trifolium pratense*): menopausal symptoms.   *Good*

St John's wort (*Hypericum perforatum*): mild to moderate   *Good*
depressive states.

Saw palmetto (*Serenoa serrulata*): benign prostate hyperplasia.   *Medium*

Tea tree (*Melaleuca alternifolia*): fungal infections.   *Medium*

Thyme (*Thymus vulgaris*): bronchitis.   *Poor*

Valerian (*Valeriana officinalis*): insomnia.   *Medium*

Willow (*Salix alba*): pain.   *Medium*

The second important point about Table 1 is that it is not comprehensive. Although the table includes more than thirty herbal medicines, we were forced to omit many remedies simply because they have not been properly tested. And without decent trials it is impossible to give an indication of whether or not a particular treatment is effective for a particular condition. If a herbal remedy does not appear in the table, then it is probably safe to assume that there is no convincing evidence to support its use.

The third point also relates to an omission, because the table makes no reference to the efficacy of so-called *individualized herbal medicines*. These special herbal mixtures are not bought over the counter, but rather they are concocted by a traditional herbalist after a detailed personal consultation. Traditional Chinese healers, Ayurvedic healers and European traditional herbalists usually practise this form of individualized herbal medicine, combining several herbs in order to find the mixture that is most suitable for the characteristics of an individual patient. It may depend on the patient's history, background, personality and environment, as well as on the current symptoms. This means that two patients presenting the same symptoms may receive very different herbal mixtures. It is harder to test this form of herbal medicine because of its individualized nature, but it is certainly not impossible. Indeed, there have been several high-quality randomized clinical trials.

Typically these trials involved dividing a group of patients with a particular condition, such as irritable bowel syndrome, into three sub-groups. Group A would receive a standard herbal treatment appropriate to the condition, such as peppermint, while groups B and C would be seen by a highly experienced herbalist who created an individualized remedy for each patient. Patients in group B would then receive their own personal remedy, while group C would receive a placebo remedy that looked and tasted similar to the individualized remedies, but which was inactive. Although patients in group A realized that they were receiving a standard herbal remedy, those in groups B and C were not aware of whether they were receiving an individualized remedy or the placebo. In general, the results of these studies are disappointing, because individualized herbal remedies either failed to perform better than placebo, or failed to perform better than the standard herbal remedy. Hence, our advice would be to avoid individualized herbal remedies – at worst they are an expensive placebo and at best an expensive option compared to straightforward herbal medicines, such as peppermint bought over the counter.

The fourth and final issue about herbal medicine – both those that appear in Table 1 and those that do not – is safety. As discussed in the previous chapter, patients need to know if an alternative medicine is both effective and safe. Arguably, safety is even more important than effectiveness.

First, do no harm

'First, do no harm' is not, as many people assume, part of the Hippocratic Oath. Nevertheless, Hippocrates did ascribe to this belief, and expressed very similar advice to doctors in his text *Of the Epidemics*: 'As to diseases, make a habit of two things — to help, or at least to do no harm.'

Modern medicine interprets this edict in terms of benefit versus risk, because we now accept that almost every medical intervention carries a risk of side-effects. Hence, before embarking on any treatment regime the doctor and patient should agree that the likelihood and extent of the potential benefit outweigh the risk and severity of adverse side-effects. So far, we have looked only at the possible benefits associated with some herbal medicines, but now it is time to examine the possible risks.

It is important to remember that the majority of powerful chemicals found in plants, those that may help deal with human disease, have evolved to serve a very different purpose. For example, some of these chemicals will have evolved in order to protect the plant from insects, and if these natural insecticides can poison bugs then it is highly likely that they will, in sufficiently high doses, harm humans too.

We will begin by discussing the drawbacks of St John's wort, because, as we have seen, it is one of the most popular and effective herbal remedies currently on the market. The main concern with St John's wort is that it contains chemicals that can interfere with other drugs that a patient might be taking. In fact, St John's wort can inhibit the impact of over half of prescribed medicines, including some anti-HIV and anti-cancer drugs. This is because St John's wort stimulates enzymes in the liver that destroy other drugs before they can do their job. Moreover, St John's wort reduces the activity of a transport mechanism that would otherwise carry drugs from the gut into the bloodstream. Essentially, this herbal remedy can serve a double whammy to other drugs, either by destroying them or by blocking their delivery.

Authorities in both Sweden and the UK have advised women using oral contraceptives not to take St John's wort, as there are several cases which indicate that the herb inhibits the normal action of contraceptives and thereby could lead to pregnancy. Similarly, concerns have been raised over kidney transplant patients, because St John's wort interferes with the action of *cyclosporine*, an immunosuppressant drug that helps prevent organ rejection. In one case, a twenty-nine-year-old woman in Arkansas began taking St John's wort for depression while also taking cyclosporine in the wake of a kidney and pancreas transplant. Her transplant had been successful, but the cyclosporine levels in her blood fell and both her kidney and pancreas function began to decline. Her doctors remained perplexed for several weeks, because the patient did not bother to tell them that she was taking St John's wort. When this issue came to light, she was asked to halt her intake of St John's wort and the doctors tried to boost her cyclosporine levels. Unfortunately, it was too late – the kidney was rejected and the patient had to return to a programme of dialysis.

The problems caused by herbal remedies interfering with conventional medicines are partly due to a lack of awareness among the general public that herbal medicines carry risks. A large proportion of the public assumes that herbal remedies are inherently safe because they are natural. An Israeli survey, for instance, revealed that 56 per cent of people using herbal remedies believed that 'they caused no side-effects'. This helps to explain the results of a survey of 318 outpatients being treated for cancer at the Royal Marsden Hospital in London – 52 per cent of them were using alternative supplements, but less than half of these patients had bothered to inform the doctors and nurses who were treating them.

Even if a patient is not taking any other medication, St John's wort can still cause problems. A 1998 review linked the remedy to several types of adverse reactions, such as gastro-intestinal symptoms, dizziness, confusion, tiredness, sedation and dryness of the mouth. However, it is important to stress that these sorts of adverse reactions are only possibilities, and the risk might be considered acceptable if a patient receives sufficient benefit from using St John's wort. Indeed, it is generally accepted that this particular remedy has fewer and less serious adverse effects than some conventional anti-depressant drugs. Hence, St John's wort can be a useful herbal remedy, as long as the patient is aware of its inherent problems, as long as there is no interference with any conventional drugs being taken, and as long as the patient's GP is kept informed.

Unfortunately, with some other herbal remedies, the adverse

effects are more serious and certainly outweigh any benefit. In the early 1990s, a Belgian doctor called Jean-Louis Vanherweghem was puzzled by the appearance of two young women at his clinic; both had suffered sudden and unexplained kidney damage, otherwise known as *nephropathy*. After a bit of questioning, he learned that both women had been following the same slimming regimen, which involved the use of various Chinese herbs. The link between the herbs and the kidney failure was only a hunch at this stage, but it was confirmed when local records showed that seven other women under fifty years of age had suffered similar kidney failure in 1991 and 1992 and all of them had followed the same herbal slimming programme.

Vanherweghem published his observations in the *Lancet* in 1993, and within a year he published a follow-up paper which identified seventy cases of what was becoming known as Chinese herb nephropathy. Thirty of these cases had been fatal. Eventually, after examining and testing the mixture of herbs that was common to all the cases of kidney failure, it became apparent that the culprit was a herb known as *aristolochia*.

Further concerns were raised in the late 1990s when the herb was also linked to cancer. Belgian doctors discovered that 40 per cent of patients diagnosed with Chinese herb nephropathy also showed signs of multiple tumours. Although this was enough evidence for several countries around the world to ban the sale of products containing aristolochia, some herbal practitioners and manufacturers still maintained that it was a safe plant and that something else must be responsible for the kidney failure and tumours. After all, aristolochia had been used for centuries and there had never been any previous indications that it might be toxic.

Indeed, the ancient Greeks, Romans, Chinese and Native Americans have all relied on aristolochia to treat everything from snake bites to headaches. Because the curved shape of the plant resembled the birth canal, European herbalists particularly encouraged its use to ease labour and induce menstruation – hence its other name, 'birthwort'. We now know, however, that all these patients were being gradually poisoned. The reason that traditional healers would not have noticed the link between the herb and subsequent kidney failure would have been because the onset of nephropathy would have been several months or years later.

The dangers of aristolochia are discussed by the investigative journalist Dan Hurley in his book *Natural Causes*, which also reveals the dangers of many other herbal medicines. One of the most recent examples in his catalogue of horrors is *ephedra*, a remedy extracted from the Chinese plant ma huang (*ephedra sinica*). Scientists had long been concerned about the side-effects of ephedra, so they developed a safer version called *pseudoephedrine*, which acts as an effective decongestant and which can still be bought today as a component in many cold remedies. Nevertheless, the original herbal extract has continued to be used by millions of people, particularly athletes and slimmers, to improve physique and lose weight. However, by 2005 there was strong evidence that 19,000 people had suffered severe reactions and at least 164 had died as a result of using ephedra. The most prominent case was Steve Bechler, a pitcher for the Baltimore Orioles, who died in 2003 during training as a result of heat-stroke. Ephedra increases perspiration and dehydration, which is why the medical examiner concluded that ephedra had played a 'significant role' in Bechler's sudden death. The sale of ephedra is now banned in most countries, although it is still readily available via the internet.

As well as the dangerous adverse effects associated with several herbal remedies, there is also another serious risk, namely the problem of contamination. In 1999 Jerry Oliveras, a food and medicines quality inspector, testified in front of an American Federal Drug Administration panel:

> Botanicals coming out of the People's Republic of China have everything from no real detectable levels of heavy metal to just about every heavy metal you want to think about. We have products coming into this country that are predominantly cinnabar. Not just cinnabar, which is a mercury salt, but also cinnabar heavily contaminated with soluble lead salt. They are sold over the counter. Go down to Chinatown, get some little red pills and take them, and go about your happy way while you're slowly poisoning yourself to death.

Ayurvedic herbal medicines are equally prone to heavy metal contamination. In 2003, a group of Boston medical researchers trawled their local shops and purchased seventy distinct Ayurvedic herbal medicine products. One in ten contained more arsenic than the standard safety level, with the worst case having an arsenic content 200 times greater than the allowed level. One in ten products also contained excess mercury, with the worst case having over 1,000 times more mercury than the recommended safety level. Most worrying of all, one in five products also contained excess lead, with the worst case having over 10,000 times more lead than the recommended safety level.

Sometimes the contaminants in herbal medicines are not toxic metals, but rather conventional pharmaceuticals, deliberately introduced in order to obtain the desired effect. For example, in 1998 the herbal sedative Sleeping Buddha was found to contain the conventional sedative drug *estazolam*, and five Chinese herbal diabetes products were tested in 2000 and found to contain the diabetes drugs *glyburide* and *phenformin*. Perhaps the most common contaminants are corticosteroids, which are added to herbal eczema creams, and

Viagra, which is introduced into some herbal aphrodisiacs to give the desired effect.

We already know from Table 1 that there is a lack of evidence to support the claims made for several herbal products, which means that many of them may be ineffective. But concealing pharmaceuticals within these ineffective herbal products leads to a dream scenario for the manufacturers and retailers. The product is still viewed as natural and at the same time it is very likely to be effective. There are, however, serious problems with this deceitful practice and the dream scenario can easily turn into a nightmare. Apart from the legal and ethical issues, the patients are unwittingly consuming a drug, thereby exposing themselves to an unknown hazard. The drug might interfere with other medicines being taken, causing adverse reactions. Alternatively, a patient might turn to a herbal remedy because he or she is allergic to a particular pharmaceutical, but if the herbal remedy is contaminated with that selfsame pharmaceutical, then the patient is duped into taking the very thing that he or she is trying to avoid.

The most infamous case of herbal medicine contamination concerned PC-SPES, a remedy that was supposedly based on a mix of Chinese herbs. It was promoted as being beneficial to prostate health and as a treatment for prostate cancer – PC is an abbreviation of prostate cancer and SPES is Latin for *hope*. Men began using it in the mid-1990s as an apparently safe and natural alternative to hormone treatment. By 2001, however, it became clear that PC-SPES had been doubly contaminated. The first contaminant was *diethylstilbestrol*, an artificial substitute for oestrogen, which had fallen out of favour in the 1970s due to its numerous adverse reactions, including blood clots. In hindsight, this explained both the effectiveness of PC-SPES and the fact that some users had died of thrombosis.

The second contaminant was *warfarin*, a blood-thinning agent used both in medicine and as a rat poison. This had presumably been introduced to counter the adverse effects of the artificial oestrogen. Unfortunately, the addition of warfarin caused other problems, namely excessive bleeding. A sixty-two-year-old man who used PC-SPES to combat prostate cancer arrived at a hospital in Seattle with uncontrolled bleeding. According to Dr R. Bruce Montgomery, one of the doctors who documented the case, 'He developed spontaneous bleeding from many places. He had a rapid heartbeat from the large amount of bleeding and low blood pressure.'

So far, we have focused on how the herbal-medicine industry can harm humans, and we will return to this issue shortly. First, however, it is worth briefly noting that the industry in natural remedies can also damage nature itself. This may come as a surprise to some users of herbal treatments, but the harvesting of wild plants for medicines is a genuine threat to the survival of some species. According to Chen Shilin of China's Institute of Medicinal Plant Development, 3,000 of the country's threatened plant species are used in traditional medicine. This number is in line with a study by Alan Hamilton from the World Wildlife Fund, who estimated that between 4,000 and 10,000 medical plants are under threat due to harvesting from the wild.

For example, the herb goldenseal was already under threat due to the destruction of its hardwood forest habitats, but its reputation as a treatment for a variety of conditions has led to widespread harvesting and has placed it at even greater risk. The absurdity is that there is no good evidence that goldenseal is effective for anything at all. By contrast, echinacea is not under threat because it is being cultivated. But it has a similar appearance to the Tennessee purple coneflower and smooth coneflower, which are endangered species and which are often gathered by mistake. One journalist called this the herbal-medicine equivalent of 'collateral damage'.

Some traditional herbal healers also offer remedies that

contain animal products, such as tiger bone or rhino horn, and in these cases the trade is pushing species to the brink of extinction. It is ironic that those who seek natural herbal cures often do so because they have a love of nature, and yet their desire to be at one with nature might be destroying it.

Before ending this section, we will return to the subject of how herbal medicines can harm humans and summarize some of the key issues. In particular, we will offer some important advice to help you protect yourself against the three potential dangers posed by herbal medicines:

1   Direct toxicity from the herbal medicine.

2   Indirect reactions caused by interactions with other medicines.

3   The risk due to contaminants and adulterants.

Before embarking on taking a particular herbal remedy, it is crucial that you reassure yourself that it is safe. To help you do this, we have drawn up Table 2, which shows the main risks associated with the most popular herbal medicines. Unfortunately, we cannot offer you a complete guide to the dangers of such remedies, as the list would run to dozens of pages. Moreover, new risks are discovered virtually every month. For example, in 2007 the *New England Journal of Medicine* reported the cases of three boys who developed breast tissue after their mothers had rubbed lavender or tea tree oil products on their chests. It seems that lavender and tea tree oils can mimic female hormones and inhibit male hormones, thereby giving rise to this condition.

Having looked at Tables 1 and 2 and other sources of reliable information, you may feel that a particular herbal medicine could help you, because it appears to be both relatively safe and reasonably effective. However, you still need to consider whether the remedy in question is safer and more effective than the conventional drug option. There is no point in taking a herbal medicine if there is a safer and more effective conventional treatment, especially bearing in mind that conventional drugs have generally undergone a higher degree of testing for both safety and efficacy. If the herbal remedy remains your preferred option, then we would urge you take on board the following points before embarking on your treatment:

1   Obtain your herbal medicine from a mainstream pharmacy, where you are likely to find the highest-quality products, which are probably free of contaminants and adulterants. You are also more likely to receive responsible advice regarding your particular condition and its treatment.

2   Take herbal medicines in pill form, rather than in the form of a powdered leaf, tea or concoction obtained from a herbalist. This is the best way to have some confidence that you are receiving the correct dosage.

3   Do not take individualized herbal mixtures from a traditional herbal practitioner. They could be contaminated or adulterated. Also, the more herbs you take, the higher the likelihood of adverse effects. Moreover, there is no evidence that the approach of individualizing herbal medicines is effective.

4   Be particularly careful about using herbal medicines if you are pregnant or if the treatment is intended for a child or an elderly person.

5   If you are already taking a conventional drug, then be

aware that there is the risk of interactions between the conventional drug and your herbal remedy.

6   Inform your GP and your entire healthcare team about any decision to take herbal medicines.

7   Last, but certainly not least, under no circumstances abandon your conventional medicine unless you have first discussed this at length with your GP.

This final point is crucial. Perhaps the most dangerous aspect of herbal medicines is that they often displace effective conventional medicines. If ineffective herbs replace an effective conventional treatment, then it is almost inevitable that the patient's condition will deteriorate. Worse still, if the patient is no longer being seen by a specialist in conventional medicine, then this deterioration might not be halted before it is too late.

For example, cancer patients are often confronted by the intimidating prospects of surgery, radiation and chemotherapy, sometimes referred to as 'slash, burn and poison' by critics of conventional medicine. Hence, the offer of a herbal medicine is tempting, because it is often marketed as a natural alternative, which is both safer and more effective. The key issue, however, is whether or not the herbal alternative is really safer and more effective.

One such natural treatment promoted as an anti-cancer agent is *laetrile*. It is an extract derived from various natural sources, often apricot pits, and it has been in use since the nineteenth century. Early proponents argued that laetrile could attack tumours by entering cancer cells, where it decomposed into cyanide, thus destroying the cells. Another hypothesis was that laetrile was a vitamin (even though it is not), and that cancers were due to a laetrile deficiency. Few doctors,

Table 2 – The risks of herbal medicines

This table relates to all the herbs in Table 1. Unlike conventional pharmaceuticals, herbal remedies have not been properly tested or monitored for safety, so it is impossible to assess their risks fully. Because of the lack of proper safety testing, some of the risks below are based on just one or two case reports. It is also important to note that many herbs can trigger allergic reactions. We have not included these in the table as there is insufficient space.

Aloe vera as a juice can cause diarrhoea, damage to the kidneys or electrolyte depletion. It can also interact with antidiabetic and heart medication. The gel is applied externally and is not known to cause adverse effects.

Andrographis interacts with some synthetic drugs, including antidiabetics and anticoagulants. It also might cause unwanted abortion.

Artichoke is not known to have adverse effects apart from flatulence.

Bilberry may cause blood sugar levels to drop dangerously or enhance antidiabetic medications. It can also interact with anticoagulants.

Black cohosh has been associated with about 70 cases of liver damage. It might also interact with heart medications.

Chamomile might interact with anticoagulants.

Cranberry has been associated with a rare case of thrombocyctopenia, a condition characterized by a low platelet count, resulting in bleeding.

Devil's claw has been linked to interactions with drugs such as anticoagulants and heart medications. It has also been associated with unwanted abortion.

Echinacea has been linked with asthma and rare conditions such as erythema nodosum.

Evening primrose could trigger an epileptic fit and might interact with drugs lowering blood pressure or heart medications.

Feverfew might interact with anticoagulants; can cause mouth to swell.

Garlic might cause blood sugar levels to drop. It can also exaggerate the effects of anticoagulants, and might interact with other drugs.

Ginger may cause bleeding and may interact with blood pressure drugs.

Ginkgo may cause bleeding or enhance anticoagulants; also linked with epileptic seizures and Stevens-Johnson syndrome.

Ginseng, both Asian and Siberian, might interact with anticoagulants and other drugs. Asian ginseng is also linked with insomnia, headache, diarrhoea, hypertension, mania and cardiovascular and endrocrine disorders.

Grape seed might interact with anticoagulants.

Hawthorn can amplify the effects of blood pressure and heart medications.

Hops might interact with contraceptive pill.

Horse chestnut could interact with anticoagulants and antidiabetic drugs.

Kava associated with skin problems and 80 cases of liver damage.

Lavender has caused nausea, vomiting, headache and chills. In rare cases, it might cause hormonal side-effects such as swelling of the breast tissue.

Ma huang contains ephedrine, which stimulates the nervous and cardiovascular systems and can cause hypertension, myocardial infarction and stroke.

Milk thistle has been associated with colic, diarrhoea, vomiting and fainting. It also interacts with antidiabetic and anti-viral drugs.

Mistletoe might interact with anticoagulants and other drugs.

Nettle has been associated with a rare condition called Reye's syndrome and it might interact with drugs for lowering blood pressure.

Passion flower may affect brain activity and EEG tests.

Peppermint might interact with blood pressure and heart disease drugs.

Red clover has been associated with bleeding and may interact with anticoagulants, the contraceptive pill and other drugs.

St Johns wort and its risks are discussed on pages 250–251.

Saw palmetto might affect blood platelets, which may cause bleeding.

Tea tree might cause swelling of breast tissue in rare cases.

Thyme can cause nausea, vomiting, diarrhoea, headache and other problems.

Valerian has been associated with isolated reports of liver damage.

Willow has been linked with isolated reports of liver damage and bleeding.

however, took laetrile seriously and it remained on the fringes of medicine until the early 1970s, when it was cleverly promoted and marketed, thus persuading many cancer patients to see it as their only hope for survival.

Dr Wallace Sampson, who is now editor-in-chief of the *Scientific Review of Alternative Medicine*, was a cancer specialist in California at the time, and he became puzzled when three of his patients suddenly stopped appearing at his clinic. After a little investigating he learned that they had started visiting a cancer clinic in Tijuana where they had been receiving laetrile. All of them claimed to be experiencing remarkable recoveries, but within a few months they had all died. Sampson did not dismiss laetrile immediately, but instead interviewed thirty-three other patients who were taking this treatment and compared them with twelve of his own patients. After matching them for age, sex and type of cancer, he noted that on average the patients on laetrile died sooner than the patients receiving conventional treatment.

The American Cancer Society labelled laetrile as 'quackery' in 1974 and the therapy was discouraged in America. Nevertheless, many patients still sought out laetrile and travelled to clinics in Mexico, where doctors such as Ernesto Contreras were making grand claims and even grander profits. By 1979, Contreras boasted that he had treated 26,000 cancer cases, but his achievements appeared hollow when the Federal Drug Administration asked for details of twelve of his most impressive cases. It turned out that six patients had died of cancer and one still had cancer, two had turned to conventional therapy and three could not be located. Nevertheless, patients continued to flock to Mexico, including Steve McQueen, who died in 1980 within five months of turning to laetrile.

At last, in 1982, the *New England Journal of Medicine* published a paper which stated conclusively that laetrile was ineffective. Four prominent cancer clinics monitored 178 cancer patients taking laetrile and observed that their overall condition deteriorated in the way that would be expected if they were receiving no treatment at all. Worse still, the researchers suspected that patients might be poisoned by laetrile (also known as *amygdalin*): 'Patients exposed to this agent should be instructed about the danger of cyanide poisoning, and their blood cyanide levels should be carefully monitored. Amygdalin (Laetrile) is a toxic drug that is not effective as a cancer treatment.' An accompanying editorial noted: 'Laetrile has had its day in court. The evidence, beyond reasonable doubt, is that it doesn't benefit patients with advanced cancer, and there is no reason to believe that it would be any more effective in the earlier stages of the disease . . . The time has come to close the books.'

Despite such evidence, far too many patients continue to spurn conventional treatment in favour of laetrile and other plant-based medicines. The ultimate result is simply a lower survival rate. If academic studies are not enough to make this point, then perhaps the appalling case of Joseph Hofbauer might serve as a warning. Aged eight and suffering from Hodgkin's disease, Joseph was taken off conventional treatment and instead was given laetrile. The New York State authorities had attempted to prevent Joseph's parents from following this path, but a family court judge ruled against the State. Chemotherapy would have offered Joseph a 95 per cent chance of surviving five years and becoming a teenager. Laetrile treatment, however, meant that he was dead within two years.

An equally tragic case, this time concerning St John's wort, also demonstrates what can go wrong when patients follow the path of herbal medicine and ignore the benefits of conventional medicine. The patient in question was Charlene Dorcy, a Canadian woman who had suffered severe depression and had attempted suicide several times since the age of thirteen. As an adult, she was diagnosed with paranoid schizophrenia. In the mid-1990s, however, she seemed to have turned the corner thanks to the mood-stabilizing drug *Tegretol*. Indeed, Charlene was described as a 'success story' when she was interviewed by the *Columbian* newspaper as part of an article on the stigma associated with mental illness.

Although Tegretol is known to have several potential side-effects, these are well understood and in Charlene's particular case there were no signs of any significant problems. Nevertheless, Charlene became convinced that a natural alternative would be preferable to Tegretol and she switched to St John's wort soon after her newspaper interview. We already know that St John's wort can give rise to adverse side-effects and interact with other medications, but an equally dangerous hazard arises when it is used to treat an inappropriate condition. And Charlene's condition was wholly inappropriate for treatment by St John's wort.

St John's wort can be effective for mild or moderate depression, but it does not seem to help with severe depression or other forms of mental illness. Worryingly, Charlene was not atypical when she used St John's wort in an inappropriate manner, because a survey of 30,000 Americans published in 2007 found that the majority of those people who self-administer herbal remedies do so in a way that runs counter to the evidence.

When Charlene switched medication, this had the double effect of depriving her of the benefits of Tegretol and exposing her to the adverse effects of St John's wort, which has been associated with aggravating psychosis in schizophrenics. Soon afterwards her condition deteriorated, and she exhibited volatility, intolerance, mood swings and repeatedly attempted suicide.

On 12 June 2004, after weeks of particularly erratic behaviour and further suicide attempts, Charlene Dorcy drove her two children to an abandoned quarry. She made two-year-old Brittney and four-year-old Jessica sit on the ground and then shot them with a .22-calibre rifle. She drove back to Vancouver, called the police and then led detectives back to the quarry where they found the children's bodies.

Why do smart people believe odd things?

We have drawn upon the results of hundreds of scientific papers in order to examine the four major strands of alternative medicine: acupuncture, homeopathy, chiropractic therapy and herbal medicine. While there is tentative evidence that acupuncture might be effective for some forms of pain relief and nausea, it fails to deliver any medical benefit in any other situation and its underlying concepts are meaningless. With respect to homeopathy, the evidence points towards a bogus industry that offers patients nothing more than a fantasy. Chiropractors, on the other hand, might compete with physiotherapists in terms of treating some back problems, but all their other claims are beyond belief and can carry a range of significant risks. Herbal medicine undoubtedly offers some interesting remedies, but they are significantly outnumbered by the unproven, disproven and downright dangerous herbal medicines on the market.

In general, the global multi-billion-pound alternative medicine industry is failing to deliver the sort of health benefits that it claims to offer. Therefore millions of patients are wasting their money and risking their health by turning towards a snake-oil industry. Also, bear in mind that we have concentrated our book on the more respectable end of the alternative medicine industry. It is shocking to think that there are dozens of even more dubious alternative therapies, which make even more outlandish claims in order to extract even more money from their patients.

These wacky therapies are among those included in the appendix, where we devote one section to each of over thirty forms of treatment. We examine each therapy's history, practice, claims and dangers. A few, such as yoga, do seem to offer genuine medical benefit, but the majority are unproven or disproven.

For example, magnet therapy is at the more negative end of the spectrum. Healers throughout the centuries have claimed that magnets have curative powers. Cleopatra supposedly wore a magnet to preserve her youth, the sixteenth-century Swiss physician Paracelsus declared that 'all inflammations and many diseases can be cured by magnetism', and in 1866 Dr C. J. Thacher's catalogue offered a complete body suit with 700 magnets, which provided 'full and complete protection of all the vital organs of the body'. Today the annual global market for therapeutic magnets is over $1 billion, which includes magnetic bracelets, shoe insoles, neck braces and even pillows. The manufacturers boast that magnets placed close to the body can treat various ailments, such as helping to heal bones, improving blood flow and relieving pain. Unfortunately, the rigorous research conducted on magnet therapy does not back up any of these claims. Magnet therapy would not be a very serious issue if it were merely a matter of arthritis sufferers wasting £10 on a useless magnetic bracelet, but the problem extends to dozens of websites offering products costing up to £2,500, including mattresses that can supposedly treat cancer.

A quick internet search will reveal crystal healers, reflexologists, aura cleansers and all sorts of other peculiar practitioners who make ambitious claims with no scientific evidence to back them up. For instance, when we searched Google, the first link took us to a clinic offering *tachyon therapy*, which can apparently heal broken bones and torn ligaments. Tachyons are particles that can travel faster than the speed of light and were hypothesized half a century ago by physicists. Bearing in mind that nobody has yet proved their existence, it is surprising to learn that somebody has been able to exploit them for medical purposes! Moreover, this clinic also offers an even more bizarre and sensational therapy: 'Multi-dimensional DNA Surgery is a channelled technique for clearing dysfunctional patterns at the level of the DNA and replacing them with Divine qualities'.

Many of these websites contain buzzwords such as energy, waves and resonance. These words certainly have scientific significance when used appropriately, but they are largely meaningless when used in the context of alternative medicine. For example, therapeutic touch is a form of alternative medicine that works by supposedly manipulating a patient's 'energy fields' to treat a range of conditions, including pain relief, healing wounds and cancer. The practitioner usually does not even need to touch the patient, which is why the treatment is also known as 'non-contact therapeutic touch' or 'distance healing'. Therapeutic touch has much in common with *reiki therapy*, inasmuch as energy fields are supposedly manipulated, often without needing to touch the patient. Although therapists charge up to £100 for a single session of therapeutic touch or reiki, it is worth noting that nobody has ever properly defined what they mean by these human energy fields, demonstrated that they actually exist or proved that they can be manipulated to improve health.

In fact, there is plenty of evidence that such human energy fields are just a myth. In 1996 a scientist based in Colorado decided to investigate therapeutic touch by testing the abilities of twenty-one healers. Emily Rosa simply asked each healer to place both hands through two holes in a screen. She would then flip a coin to decide whether she would place her own hand close to the healer's left or right hand. The healer then had to sense Emily Rosa's energy field to decide where she had placed her hand. The twenty-one healers had 280 attempts in total and beforehand they were confident that they would consistently be able to sense the location of the scientist's hand. Chance would predict a 50 per cent success rate, but in fact the therapeutic touch healers could guess correctly only 44 per cent of the time. The experiment showed that the energy field was probably nothing more than a figment in the imaginations of the healers.

At this point, it is worth pointing out that Emily was only nine years old when she conducted this experiment. It was originally her school science fair project, but two years later she wrote up her research with the help of her mother, a nurse, and it was published in the highly respected *Journal of the American Medical Association*, making Emily the youngest person (as far as we know) ever to have a research paper published in a peer-reviewed medical journal. Not surprisingly, there were some critics who remained unimpressed with Emily's paper, which was entitled 'A Close Look at Therapeutic Touch'. Dolores Krieger, who formulated the principles of the therapy, accused the research of being 'poor in terms of design and methodology'. In fact, Emily's protocol was simple and clear and her conclusion was hard to fault. Moreover, nobody has ever come up with an experiment that has overturned her findings.

According to Emily's research and other trials, therapeutic touch, reiki and many other related therapies are based on nothing more than wishful thinking. Any benefit that they offer seems to be entirely attributable to the placebo effect. Nevertheless, these therapies are part of a massive, global industry – according to Emily's paper there are 100,000 trained therapeutic touch healers worldwide, presumably treating millions of patients, and earning several £100 million every year. The typical patients for these so-called energy therapies and other ineffective alternative therapies are neither stupid nor naïve. This raises an interesting question – why is it that a nine-year-old child was capable of testing and disproving the claims of therapeutic touch, while grown adults are completely fooled by these healers?

This section is devoted to looking at the reasons why smart people believe in alternative medicine, when we have shown that so much of it is ineffective. The reasons must be compelling in order to persuade millions of people to part with billions of pounds in a misguided attempt to protect their most precious asset, namely their health.

The initial reasons why people find alternative medicine appealing are often related to the three core principles that underlie so many of the therapies – they are said to be based on a more *natural*, *traditional* and *holistic* approach to healthcare. Advocates of alternative medicine repeatedly cite these principles as strong grounds for adopting alternative medicine, but, in fact, it is easy to show that they are nothing more than clever and misleading marketing ploys. The three principles of alternative medicine are really fallacies:

1   **'Natural' fallacy**
     Just because something is natural it does not mean that it is good, and just because something is unnatural it does not mean that it is bad. Arsenic, cobra poison, nuclear radiation, carthquakes and the ebola virus can all be found in nature, whereas vaccines, spectacles and artificial hips are all man-made. Or, as the *Medical Monitor* put it, 'Nature has no bias and can be seen at work as clearly, and as inexorably, in the spread of an epidemic as in the birth of a healthy baby.'

2   **'Traditional' fallacy**
     The notion that traditional is a good quality helps many alternative therapists because it means that the placebo

effect is reinforced by a dose of nostalgia. However, it would be wrong to assume that traditional therapies are inherently good. Bloodletting was traditional for centuries, and throughout this time it harmed many more people than it healed. Our job in the twenty-first century is to test what our ancestors have bequeathed us. In this way, we can continue with the good traditions, adapt the traditions with potential and abandon those traditions that are mad, bad or dangerous.

3   'Holistic' fallacy

Alternative therapists use the term holistic to imply that their approach is superior to conventional medicine, but this 'more holistic than thou' attitude is unjustified. Holistic merely means taking a whole-person approach to medicine, and conventional doctors will also treat their patients holistically. GPs consider a patient's lifestyle, diet, age, family history, medical background, genetic information and the results from a variety of tests. If anything, conventional medicine takes a more holistic approach than alternative medicine. This was demonstrated in Chapter 3 when we compared conventional against homeopathic healthcare in the case of a student looking for advice about malaria prevention. The conventional clinic offered a long consultation, covering not just the drug options, but also the use of insect repellent, appropriate clothing and the student's medical history. By contrast, the majority of homeopaths offered a very short consultation and gave no advice on basics such as bite prevention.

In addition to promoting their own fallacious, yet superficially attractive, core principles, the alternative-health industry also tries to recruit patients by condemning mainstream scientists. Alternative therapists are, of course, aware that scientists are largely critical of alternative treatments, so they attempt to undermine the scientific criticisms by questioning the credibility of science itself. The attacks on science cover three areas, but again we can see that the alternative therapists are basing their propaganda on fallacies:

1   **'Science cannot test alternative medicine' fallacy**

As we have shown throughout this book, science is more than capable of testing alternative medicine. Indeed, that is exactly why scientists are sceptical about its many and varied claims. All these alternative therapies boast that they offer real and significant physiological impacts, ranging from pain relief to curing cancer, and medical science has developed techniques for measuring all these medical outcomes. If science cannot detect the alleged benefits of alternative medicine, then it is either because they do not exist or because they are too small to be worth bothering about.

2   **'Science does not understand alternative medicine' fallacy**

This is true, but irrelevant. Failing to understand how a therapy works has never been a barrier to accepting that it does work. Indeed, the history of medicine is littered with breakthrough treatments that were clearly effective and yet not initially understood. For example, when James Lind discovered that lemons could prevent scurvy in the eighteenth century, he did not understand how the lemons actually worked. Nevertheless, his treatment spread across the world. It was only in around 1930 that scientists isolated vitamin C and understood why lemons safeguard against scurvy. If a particular alternative treatment were proved to be effective tomorrow, then scientists would accept it and immediately attempt both to apply it and to understand its underlying mechanism.

3   **'Science is biased against alternative ideas' fallacy**
This is even more absurd than the first two fallacies. People who have alternative ideas are mavericks, and the whole of modern science has been built by mavericks, from Galileo right through to the latest crop of Nobel Laureates. In fact, it could easily be argued that all great scientists are mavericks in some way. Unfortunately, the converse is not true – all mavericks are not necessarily great scientists. Having come up with a radical idea, the challenge for any maverick is to prove to the rest of the world that the idea is correct, but this is where the majority of pioneers of alternative medicine come unstuck.

The last fallacy is worth exploring further, as science is often portrayed as a closed shop, when in fact the scientific community lovingly embraces those mavericks who can find evidence to support their claims. For example, in the 1980s, Australian researchers Barry Marshall and Robin Warren suggested that most peptic ulcers are caused by bacteria. The conventional view was that excess acid, the wrong diet and too much stress were the major factors in causing ulcers, which is why initially nobody took Marshall and Warren's revolutionary idea seriously. However, in a famous and courageous experiment, Marshall successfully identified the rogue bacteria, cultured it, swallowed it and developed ulcers, thereby proving that ulcers had a bacterial origin. Obviously, other medical scientists were now convinced by the new theory and rewarded Marshall and Warren with a Nobel Prize in 2005. Even more importantly, a combination drug therapy has been developed to ward off the bacteria and cure those plagued by ulcers – this drug therapy is more effective, cheaper and quicker than previous treatments, so millions of people around the world have benefited from this once maverick idea.

It does not matter who the mavericks are, or how, when and where they come by their discovery. Even lucky discoveries are readily recognized by the establishment if they can be validated. Viagra, one of the most successful drug discoveries in recent years, was originally developed to treat angina, but a pilot study showed that it did little to alleviate this condition. However, when researchers decided to stop the trial early and recall any unused pills, they were perplexed by the reluctance of the trial volunteers to return them. Subsequent interviews revealed that Viagra had an unexpected and desirable side-effect. Further trials and safety tests have resulted in Viagra's current widespread availability for the treatment of impotence. No homeopathic, chiropractic, herbal or acupuncture therapy has been able to show such a dramatic impact on the treatment of erectile dysfunction.

Curiously, while alternative medicine is often quick to criticize science on the one hand, it is equally keen to use science to its own advantage whenever it is convenient. But, yet again, alternative therapists are relying on flawed arguments and faulty notions to promote themselves. These fallacies fall into three broad categories:

1   **'Scientific explanation' fallacy**
Some alternative therapists employ scientific explanations to give credence to their treatment, but just because an explanation sounds convincing, it does not mean that it is true. For example, magnet therapists sometimes argue that magnets act on the iron component in our blood to restore the body's electromagnetic balance, but this makes no scientific sense. The

haemoglobin in our blood does indeed contain iron, but it is not in a form that responds to magnetism – this can be crudely tested by placing a strong magnet next to a drop of blood. Sometimes the explanations in alternative medicine contain pseudo-scientific jargon, such as a London-based healing clinic which uses phrases such as 'the client's electromagnetic circuitry' and 'defragmentate the body'. This jargon may be impressive to a non-specialist, but it is scientifically meaningless. We, the authors of this book, have a medical doctorate and two PhDs (particle physics and blood rheology) between us, yet we are baffled by these words.

2   'Scientific gadget' fallacy

Just because some alternative therapists employ gadgets that look impressive, it does not mean that they actually work. The Aqua Detox, for instance, is an electrical footbath that claims to draw toxins out of the body. The water actually turns brown during this process, which seems like evidence that the body is being cleansed. A UK-based alternative clinic claims that this treatment 'has helped people of all ages, from babies (via a unit that fits in the bath) to the elderly, and has eased things like digestive disorders, skin conditions, chronic tiredness and migraine to name but a few . . . it's been used by cancer patients to draw radioactivity from their bodies after chemotherapy.' Unfortunately, the water in an Aqua Detox unit only turns brown because of a simple electro-chemical reaction which rusts the iron contacts on the side of the footbath. In other words, the water is not becoming saturated with toxins, it is merely awash with rust. Medical journalist Ben Goldacre analysed some water before and after an Aqua Detox session. Sure enough, the iron content in the water increased by a factor of fifty, yet there were no signs of the most obvious toxins. In one further test, Goldacre placed a Barbie doll into the footbath, and once again the water turned brown, which only reinforces the view that the discoloration was related solely to the machine's own functioning.

3   'Scientific clinical trial' fallacy
We have stressed the vital role of clinical trials in determining the truth about a treatment, but just because an alternative therapist cites a trial in support of a particular treatment, it does not mean that it is effective. The problem here is that a single trial is not enough to demonstrate that a particular therapy works, because that particular trial might have been prone to error, the vagaries of chance or even fraud. That is why we have not based the conclusions in this book on individual pieces of research, but instead we have examined the broad consensus drawn from the totality of the reliable evidence. In particular, we have relied on meta-analyses and systematic reviews, in which a team of scientists has set itself the task of examining all the research in order to come to an over-arching conclusion.

The importance of the third fallacy can be illustrated by looking at research into whether or not prayer can help patients. Scientists already accept that patients who know that their relatives are praying for them may have a slightly better chance of recovery. This can be explained by obvious psychological effects, such as the likelihood that prayers give the patient a sense of love, hope and support at a time of crisis. Therefore, there is no need to resort to a paranormal explanation for the benefit given to patients who are aware of family prayers. However, scientists have wondered what would happen to patients who are being prayed for, but who are unaware of this spiritual intervention. Any resulting benefit could not be attributed to psychological factors, because the patient is blind to the prayers. Hence, if these patients were to benefit from secret prayers, then it would indicate some level of divine intervention.

One of the most famous studies into the power of prayer was published in 2001 by three authors, including one scientist based at the prestigious Columbia University in New York. It looked into whether prayer could help patients receiving fertility treatment. The trial involved 199 women in South Korea, 100 of whom received IVF treatment and had their photographs sent to prayer groups in Canada and Australia, and 99 of whom received just IVF. Crucially, the women did not know whether or not they were among the group being prayed for, yet the women being prayed for had twice the pregnancy rate of the control group – a remarkably significant result.

The research was published in the respected *Journal of Reproductive Medicine*, and it was then reported around the world with headlines declaring that scientists had proved that prayers can help patients. Meanwhile, other researchers thought that it was premature to jump to any such conclusion. Although this was certainly an interesting piece of research, it was just a one-off trial and the scientific community is reluctant to accept the conclusion from a single piece of research, particularly when its conclusion is so extraordinary. The only way that the results from this research would be taken seriously would be if follow-up clinical trials pointed to the same conclusion. Alternatively, if subsequent studies showed no effect, then it would be safe to assume that the initial study was flawed in some way and it could then be reasonably disregarded.

In fact, there was already a similar prayer study under way in

2001. This one involved 799 patients in an American coronary care unit. Half of them unknowingly received 'intercessory prayer' from groups of healers for twenty-six weeks, and the other half received no such prayers. The number of deaths, heart attacks and other serious complications were similar in both groups, which implied that prayers were having no effect.

In another study, which took place in 2005, 329 patients under-going angiograms or other cardiac procedures received no prayers, while 371 patients received prayers from Christian, Muslim, Jewish and Buddhist prayer groups. Unfortunately, the prayers had no measurable effect on serious cardiovascular events, hospital re-admission or death. And, in 2006, the results of a ten-year study costing $2.5 million were published by researchers studying the effect of prayer on over 1,000 cardiac bypass surgery patients at six American medical centres. Christian groups prayed for half the patients for several years, while the other half received no such prayers. Again, the average outcome was the same for both groups, implying that the prayers were ineffective.

Today, the balance of evidence clearly goes against the possibility of divine healing via prayer. This means that the original power of prayer study, which gave a shockingly positive result, probably contained serious errors in the way it was conducted. In fact there are several reasons to be suspicious about that particular trial.

First, after the research was published, it became apparent that the study was conducted without informed consent. To be more specific, the women involved had no idea that their photographs were being sent to the prayer groups. Bearing in mind that infertility is a personal and sensitive issue, this is a major breach of protocol. Dr Bruce Flamm, who has investigated the study, pointed out:

Furthermore, because the study was conducted in Korea,

where the majority of the population is Buddhist, Shamanist, or nonreligious, many study patients might have objected to Christian prayers as unwanted, blasphemous, or antithetical to their personal beliefs. But since the study was conducted without their knowledge or permission, the study subjects had no way to voice their objections or to opt out of the study.

Per se, this lack of consent did not invalidate the results of the trial, but it prompted one of the three researchers to reveal another worrying issue. Rogerio Lobo, who brought credibility to the research because of his position as chairman of a department at Columbia University, admitted that he had not been involved in conducting the actual research and instead merely helped to edit and publish the research paper. Dr Lobo has now removed his name from the paper, implying that he no longer believes it to be a respectable piece of research and prefers not to be associated with it.

Daniel Wirth, who was the second author, still seems to believe that the research is credible, but his own integrity has been questioned ever since 2004 when he pleaded guilty to criminal fraud and to using numerous fake identities to commit felonies. He was sentenced to five years in federal prison. The third author of the prayer-infertility paper was Dr Kwang Cha – he holds the unique position of both remaining loyal to the research and not being a convicted felon.

There is clearly a risk that patients might be aware of this particular prayer study, without knowing about its dubious background or the existence of all the other studies that contradict its conclusion. This, in turn, could leave patients and their families with an unjustified confidence in the power of prayer, thereby tempting them to pay for and rely on spiritual healers.

This brief history of prayer-fertility research illustrates a general point about alternative medicine. Before deciding whether or not to invest time, money and hope in an alternative treatment, it is important for patients to know about the overall conclusion from all the research conducted into that particular treatment. That is why we have devoted four chapters to examining the evidence with respect to the four main alternative therapies. In our appendix, we have applied the same approach to analysing over thirty more alternative therapies – our conclusions are much more concise, but they are equally rigorous.

As you have probably realized by now, with some important exceptions our conclusions about alternative medicine are largely negative. Over and over again, we are forced to use words such as 'disproven', 'unproven', or even 'dangerous'. This is what the balance of evidence tells us, and we have done our level best to explain how we have arrived at our conclusions and why you should take them seriously. There is, however, one reason why you still might be reluctant to accept our conclusions, and why instead you might be willing to give alternative medicine the benefit of the doubt. We will cover this reason, which is both compelling and misleading, in the final section of this chapter.

Seeing is believing

For many patients, scientific evidence is not the deciding factor in whether or not they adopt a particular alternative therapy. Even if they are aware that the overall conclusion based on all the research is negative, patients are still likely to adopt a therapy if they have personally witnessed its benefits with their own eyes. After all, seeing is believing. This reaction is quite natural and wholly understandable, yet it exposes patients to the risk of ineffective and possibly dangerous treatments.

If we take homeopathy as an example, then millions of

people are convinced that it is effective because of their own personal experience – they suffer various ailments, they consume homeopathic remedies and they feel better, so it is perfectly natural to assume that the homeopathic remedy was responsible for their recovery. The fact that the scientific evidence indicates that homeopathy is wholly ineffective, as discussed in Chapter 3, carries very little weight for most people in this sort of situation.

How do we resolve this conflict between personal experience and scientific research? Two hundred years of scientific testing is unlikely to be wrong, so let us assume (for the time being, at least) that homeopathy is ineffective. This would mean that our personal experiences are somehow misleading us – but how?

The central problem is that we are tempted to assume that two events that happen one after the other must be connected. If recovery from illness takes place after taking some homeopathic pills, then isn't it obvious that the homeopathic pills caused the recovery? If there is a correlation between two events, then isn't it common sense that one event caused the other? The answer is 'No'.

We can see why a correlation should not be confused with causation if we look at a neat example invented by Bobby Henderson, author of *The Gospel of the Flying Spaghetti Monster*. He spotted a very interesting correlation between the increase in global temperature over the last two centuries and the decline in the number of pirates. If correlation is synonymous with cause and effect, then he speculated that the decline in pirates is causing global warming. Henderson therefore suggested that political leaders should encourage more pirates to take to the seas in an effort to combat global warming. This might seem ridiculous, but Henderson backed up his causal link between pirates and global warming with further evidence. For example, many people dress up as pirates for Halloween, and the months following 31 October are generally cooler than those that precede it.

Henderson's absurd pirate–climate example should be enough to show that two events happening at the same time are not necessarily linked. Hence, it is certainly conceivable that homeopathic remedies are not causing the recoveries with which they are associated. This, however, poses a new problem – how and why are patients feeling better? We can only reject homeopathy as the causal agent if we can find more reasonable explanations for why patients report improvements soon after taking homeopathic pills. As it happens, finding such explanations is relatively straightforward.

For example, the patient might be taking conventional medicine that might coincidentally take effect around the time that he or she resorts to homeopathic pills. Although it is the conventional pill that is active, the patient might credit the homeopathic pill. Another explanation is that the patient might be benefiting thanks to other recommendations from the homeopath, such as advice on relaxation, diet or exercise. These lifestyle changes can positively influence a whole range of conditions, and the benefits can easily be misattributed to homeopathic pills that are being taken at the time. We also have to consider the possibility that the homeopathic remedy is contaminated, perhaps with steroids or other conventional pharmaceuticals. In each of these cases, it is not the homeopathic pill that is helping the patient, but rather the contaminant, the homeopath's advice or the parallel conventional treatment.

Other explanations for why homeopathy seems to work rely on changes that are occurring in the patient's own body. For example, it is quite natural for symptoms to fluctuate, and it might be that the taking of a homeopathic pill coincides with an upswing in the patient's condition. Indeed, when a patient feels particularly awful, perhaps during a bout of influenza, then it might be tempting to turn to homeopathy, but at that point the only way is up. This is known as *regression to the mean*: a patient who is feeling particularly ill has probably hit rock-bottom and is very likely to start returning to their average (or mean) state.

It is also worth bearing in mind that many conditions have a limited natural duration, which means that the body heals itself given time. Unexplained lower back pain significantly improves within six weeks for roughly 90 per cent of patients who receive no treatment, so any homeopath who can retain a patient for a couple of months is highly likely to see some sort of recovery within this period. This natural recovery is likely to be inappropriately claimed as a success for alternative medicine.

Some of these explanations rely heavily on coincidence. Remarkable coincidences are rare, such as the one spotted by puzzle expert Cory Calhoun, who noticed that the letters of 'To be or not to be: that is the question; whether 'tis nobler in the mind to suffer the slings and arrows of outrageous fortune . . .' from Shakespeare's *Hamlet* can be rearranged into the highly appropriate 'In one of the Bard's best-thought-of tragedies our insistent hero, Hamlet, queries on two fronts about how life turns rotten.' On the other hand, more mundane coincidences are commonplace. With millions of people catching colds and so many of them trying alternative remedies, it is inevitable that a significant number will coincidentally experience an improvement in their condition soon after taking the remedy.

Fortunately for alternative therapists, they are in an ideal position to exploit the vagaries of coincidence and thereby take false credit for the body's own healing powers. They often treat patients with chronic conditions, which have fluctuating symptoms, providing numerous opportunities coincidentally to catch the condition on an upswing. Back pain, fatigue, headaches, insomnia, asthma, anxiety and irritable bowel syndrome are all conditions that go through unpredictable cycles of improvements and deterioration. A herbal or homeopathic pill taken when the patient is at their worst, or an acupuncture session when the patient was in any case beginning to improve, will all be perceived as the agent of change.

Even if the start of the treatment coincides with a decline in the patient's condition, then this can be excused by the so-called 'healing crisis' or 'aggravation', already discussed in Chapter 4. This is claimed to be an inherent part of many alternative therapies, whereby the patient's health is almost expected to worsen before it improves, supposedly because toxins are being ejected. In reality, this ploy merely buys the therapist more time. Eventually, when recovery actually begins, for whatever reason, the alternative therapist is still in a position to take the credit.

Many of the coincidences described so far are particularly likely to impress those patients who already have a strong belief in alternative medicine. This is because believers are vulnerable to *confirmation bias*, which is the tendency to interpret events in a way that confirms preconceptions. In other words, believers will focus on information that supports prior beliefs and ignore information that contradicts those beliefs. Practitioners are highly prone to confirmation bias, because they have vested interests, both emotional and financial, in seeing the therapy work. This sort of confirmation bias is sometimes referred to as Tolstoy syndrome, after an observation made by Leo Tolstoy:

> I know that most men, including those at ease with problems of the greatest complexity, can seldom accept the simplest and most obvious truth if it be such as would oblige them to admit the falsity of conclusions which they have proudly taught to others, and which they have woven, thread by thread, into the fabrics of their life.

There is one more good reason to explain why so many people experience some form of recovery soon after taking an alternative treatment, even when the scientific evidence suggests that the treatment is entirely ineffective. It is an explanation that you may already be expecting, because it was discussed in detail in Chapter 2 – it is the placebo effect. Remember, this is the phenomenon whereby a patient responds positively to a treatment simply because of a sincere belief that the treatment is effective. The placebo effect is a very real phenomenon, so much so that doctors have probably been aware of it since ancient times, and it has been scientific- ally studied for over half a century. It is potentially very powerful, providing everything from pain relief to boosting a patient's immune system.

The evidence outlined so far in this book suggests that the majority of alternative therapies in most cases have very little to offer aside from the placebo effect. Consequently, it would be tempting to condemn all these therapies as being worthless – but this is too simplistic, as it ignores the genuine relief that can emerge as a result of the placebo effect.

This revives an issue that was briefly mentioned at the end of Chapter 2, and which is one of the most important and con- troversial questions concerning alternative medicine. Even if alternative medicine relies largely on the placebo effect, why shouldn't alternative therapists exploit placebo to help the sick, particularly when we know it can be so powerful? In the final chapter of this book we will outline our response to this question.

6 Does the Truth Matter?

'It makes good sense to evaluate complementary and alternative therapies. For one thing, since an estimated £1.6 billion is spent each year on them, then we want value for our money.'

His Royal Highness The Prince of Wales

If you have been a particularly attentive reader, then you may have noticed that on page 5 we dedicated this book to His Royal Highness The Prince of Wales. We took this decision because the Prince has a long-standing interest in alternative medicine. In fact, as early as 1993 he established the Foundation for Integrated Health, which exists in order 'to encourage greater collaboration between conventional and complementary health practitioners and to facilitate the development of integrated healthcare.'

The Prince of Wales has spoken positively about alternative medicine on numerous occasions, whether visiting hospitals, or addressing GPs, or at the World Health Organization. He has also written several articles on the subject of alternative medicine, including one that was published in reaction to a report by the House of Lords Select Committee on Science and Technology in 2000. The Committee had concluded that many forms of alternative medicine were poorly understood because they had not been adequately tested, neither in terms of efficacy nor of safety. Responding in *The Times*, Prince Charles accepted that this was indeed the case, but he strongly emphasized another aspect of the Committee's report, namely the need for more research into alternative medicine in order to address the issues of efficacy and safety.

Under the heading 'Alternative medicine needs – and deserves – more research funding', the Prince pointed out that he had been advocating an evidence-based approach to alternative medicine for some time:

Are these therapies as good as orthodox medicine, or even in some instances better? If so, which therapies and for which conditions? In 1997 the Foundation for Integrated Medicine, of which I am the president and founder, identified research and development based on rigorous scientific evidence as one of the keys to the medical establishment's acceptance of non-conventional approaches.

Although the tone of Prince Charles's article was strongly optimistic, implying that greater evaluation would lead to greater acceptance of alternative therapies, many medical researchers were more sceptical. Either way, there was general agreement that further research was the way forward.

Since 2000, there have been some 4,000 research studies into alternative medicine published worldwide, and to a large extent this book has been written in order to answer Prince Charles's questions. He wanted more research to find out which therapies work and which do not: now that the research is on the table, we are in a position to identify those therapies that genuinely help patients, those that are pure quackery and those that lie somewhere in the middle.

The previous chapters examined the four main alternative therapies, and their conclusions reveal that these alternative treatments offer a disappointing level of benefit. At the more positive end of the scale, herbal medicine can claim a few successes, but the majority of herbs appear to be overhyped. Chiropractic therapy might offer some marginal benefit, but only for back pain – all its other claims are unsubstantiated. Similarly, acupuncturists might be able to offer some marginal benefit in terms of relieving some sorts of pain and nausea, but the effect is so borderline that there is also the strong possibility that acupuncture is worthless. And it is certain that acupuncturists are guilty of offering unproven treatments for a range of conditions, including diabetes, heart disease and infertility. Homeopathy is the worst therapy encountered so far – it is an implausible therapy that has failed to prove itself after two centuries and some 200 clinical studies.

The bottom line is that none of the above treatments is backed by the sort of evidence that would be considered impressive by the current standards of medical research. Those benefits that might exist are simply too small, too inconsistent and too contentious. Moreover, none of these alternative treatments (apart from a few herbal medicines) compare well against the conventional options for the same conditions. This dismal pattern is repeated in the appendix, where we examine many more alternative therapies.

If you find the previous paragraph somewhat blunt and disparaging, please remember that it is based on an analysis of the scientific information, which is exactly what both the House of Lords and the Prince of Wales advocated. As outlined in Chapter 1 (and indeed reiterated throughout the book), it is clear that scientific trials, observations and experiments are the fairest and best ways to establish the truth in medicine, so our conclusions cannot easily be dismissed.

In light of such disappointing results, it seems bizarre that alternative treatments are touted as though they offer marvellous benefits. In fact, not only are such treatments unproven, but over and over again we have seen that alternative medicine is also potentially dangerous. Remember, chiropractors who manipulate the neck can cause a stroke, which can be fatal. Similarly, some herbs can cause adverse reactions or can interfere with conventional drugs, thereby leading to serious harm. Acupuncture practised by an expert is probably safe, but minor bleeding is common for many patients and more serious problems include infection from re-used needles and the puncturing of major organs. Even homeopathic remedies, which of course contain no active ingredient, can be dangerous if they delay or replace a more orthodox treatment. In fact, any ineffective alternative treatment jeopardizes the health of a patient if it replaces an effective conventional treatment. This problem was clearly demonstrated by the tragic death of a Dutch comedienne called Sylvia Millecam.

Millecam shot to fame in Holland when she starred in her own TV show in the 1990s. However, in 1999 her GP noticed a small lump in her breast. He referred her to a radiologist for more tests, but the results were inconclusive. Then, instead of visiting a surgeon for further investigation, she underwent electro-acupuncture. Even when it was absolutely clear that she had breast cancer, Millecam rejected conventional medicine and visited a total of twenty-eight alternative practitioners over the course of two years. The pointless treatments that she received included homeopathy, food supplements, cell-specific cancer treatment, salt therapy and psychic healing, and her diagnoses relied on bizarre techniques such as electromagnetic and vega testing. Gradually the cancer spread and Millecam was admitted to hospital in August 2001, but it was too late. She died four days later, aged forty-five. This is appalling: if Millecam's breast cancer had been treated quickly, then she would probably still be alive. An expert medical panel looked into the case of Sylvia Millecam and concluded that she had received 'unfounded methods of treatment' and that her alternative therapists had denied her 'a reasonable chance of recovery' and caused her 'unnecessary suffering'.

And, not only do alternative therapists offer us often ineffective and sometimes dangerous treatments, they also charge us heavily for these services and products. The issue of money is problematic at every level. Parents on a limited budget might waste money on alternative medicine in a misguided attempt to improve their child's health. At the other end of the scale, national governments have much larger budgets, but these are also limited, and they also risk wasting money on alternative medicine in a similarly misguided attempt to improve the health of their nations.

Acupuncture sessions, chiropractic manipulations and homeopathic consultations can all cost upwards of £50 each and are often more than double this price. Other alternative therapists, such as spiritual healers, charge similar amounts for a session, and a full course of alternative treatment for an individual can cost hundreds or thousands of pounds. At the start of this chapter, the Prince of Wales quoted an annual UK spend of £1.6 billion in 2000, but even this is likely to be an underestimate. Surveys of the money spent on alternative medicine can give conflicting results, but the general trend has been inexorably upwards, and a recent extrapolation estimated that Britons currently spend £5 billion on alternative treat- ments – £4.5 billion spent by the public and the remaining £500 million being spent by the National Health Service. And, remember, the annual global spend on alternative medicine is estimated to be £40 billion.

One might argue that every individual has the right to spend money according to his or her own wishes, but if alternative practitioners are making unproven, disproven or vastly exaggerated claims, and if their treatments carry risks, then we are being swindled at the expense of our own good health.

In terms of UK government spending, the alternative lobby might defend the £500 million bill by pointing out that it represents less than 1 per cent of the National Health Service budget, but £500 million spent on unproven or disproven therapies could instead pay for 20,000 more nurses. Another way to grasp the impact of government spending on alternative medicine is to consider the recent refurbishment of the Royal Homoeopathic Hospital in London, which cost £20 million. The hospital is part of the University College London Hospitals NHS Foundation Trust, which had to announce a deficit of £17.4 million at the end of 2005. In other words, the overspend could have been easily reduced if money had not been spent on refurbishing a hospital that practises and promotes a bogus form of medicine.

Professor David Colqhuoun, a pharmacologist at University College, has been one of the most vocal critics of the money invested in the Royal Homoeopathic Hospital:

> It is striking that the deficit seems to be comparable with the costs of the Royal London Homoeopathic Hospital. It is true that the amount spent by the NHS on complementary medicine is not huge when compared to the entire NHS budget, but to spend anything at the same time that different parts of the trust are dismissing nurses is quite wrong.

Despite the fact that many scientists consider much of alternative medicine to be a waste of money, the Prince of Wales remains enthusiastic about its potential role within government health provision. To add weight to his views, His Royal Highness commissioned a report 'to examine evidence relating first to the effectiveness and then to the associated costs of mainstream complementary therapies'. The mainstream complementary therapies were identified as acupuncture, homeopathy, herbal medicine, chiropractic therapy and osteopathy, and the report was conducted by the economist Christopher Smallwood. When it was published in 2005, one of the main conclusions of the Smallwood Report surprised medical experts:

> Despite the fragmentary nature of the evidence, there seems good reason to believe that a number of CAM (Complementary and Alternative Medicine) treatments offer the possibility of significant savings in direct health costs, whilst others, perhaps just as expensive as their conventional

counterparts, can nonetheless deliver additional benefits to patients in a cost-effective way. In addition, the benefits to the economy of a wider application of successful complementary therapies in the key areas could run into hundreds of millions of pounds.

In light of all the negative evidence contained in our book, Smallwood's conclusion seems utterly absurd, so how did a respected economist come to such a distorted and rose-tinted view of alternative medicine? Smallwood himself admitted that he and his team had no specialist expertise in healthcare economics, and they were naïve about the research data concerning alternative medicine. This resulted in a report that contained numerous fundamental errors.

For example, Smallwood maintained that homeopathy is a cost-effective treatment for asthma, even though a Cochrane review indicates that its effectiveness has not been established. Hence, it is not so surprising that the report mistakenly claimed that £190 million would be saved if 4 per cent of British GPs were to offer homoeopathy as a front-line approach to treatment.

The Smallwood Report's conclusions were not only incorrect, but also highly dangerous. For example, if homeopathy were to be used to treat asthma, then the consequences could be disastrous, as pointed out by Richard Horton, editor of the *Lancet*:

> About 1,400 people die from asthma every year in the UK. It is a life-threatening condition that can be controlled by the effective use of drugs. The idea that homeopathy can replace conventional treatment, as the Prince's report suggests, is absolutely wrong. Not one shred of reliable evidence exists to support this incredibly misjudged claim.

Possibly more than any previous member of the Royal Family, the Prince of Wales has been outspoken on a number of issues, from architecture to opportunities for young people, from the environment to, of course, alternative medicine. In many cases, he has been a force for good, highlighting important causes and bringing them to public attention. In other cases, such as alternative medicine, he has steered the debate into sterile territory and made statements that have flown in the face of all the best expert opinion. For instance, at a healthcare conference in 2004, the Prince endorsed Gerson therapy, which is based on a severe diet regime and coffee enemas:

> I know of one patient who turned to Gerson Therapy having been told she was suffering from terminal cancer and would not survive another course of chemotherapy. Happily, seven years later, she is alive and well. So it is vital that, rather than dismissing such experiences, we should further investigate the beneficial nature of these treatments.

The Prince was thus promoting a therapy that has been discredited and which is known to be potentially harmful. Gerson therapy starves already malnourished patients, depriving them of vital nutrients. Moreover, adopting Gerson therapy often means that patients abandon their conventional treatment, thereby jeopardizing their main hope for recovery. Although the Gerson Institute has an office in California, it runs its main clinic in Tijuana, Mexico, where it claims it can cure cancer – this geographical dislocation is necessary because the US forbids doctors to practise Gerson therapy.

It was misguided at best, irresponsible at worst, for the Prince of Wales to suggest publicly that Gerson therapy might be able to treat cancer, when the evidence is to the contrary. And it would be arguably reckless for him to continue promoting alternative medicine in general when we have demonstrated in this book that very little benefit is derived from therapies such as acupuncture, homeopathy, chiropractic therapy and herbal medicine.

In short, the Prince of Wales ought to start listening to scientists rather than allowing himself to be guided by his own prejudices. Or, as Professor Michael Baum, a cancer specialist at University College, London, put it: 'The power of my authority comes with a knowledge built on 40 years of study and 25 years of active involvement in cancer research. Your power and authority rest on an accident of birth.'

Placebos – little white lies or fraudulent falsehoods?

We have shown that the majority of alternative treatments are wholly or largely ineffective in treating the majority of conditions. The term 'ineffective', however, does not mean that such remedies are of no benefit to patients, because there is always the placebo effect, which we know can offer varying levels of relief. So, should doctors encourage the use of disproven alternative treatments, which on the one hand are nothing more than fake remedies, but which on the other hand can help those patients who have sufficient faith in them? Can large parts of the alternative-medicine industry justify their existence by offering relief through belief?

Of course, patients with life-threatening conditions cannot rely on the placebo effect to rescue them, but for patients with less serious conditions the issues are more complicated. Because of this complexity, we will explore the value of placebos by focusing on homeopathy, but everything that follows is also applicable to the placebo effect in the context of other alternative therapies.

Homeopaths will argue that their remedies are genuinely

effective, but we know that the best scientific evidence concludes that homeopathic remedies are bogus and rely wholly on the placebo effect in order to benefit patients. For example, rubbing homeopathic Arnica cream on a bruise works only at a psychological level, so that a patient merely feels that a bruise is healing faster and that the pain is subsiding. Or a person with high blood pressure might take a homeopathic remedy, and the resulting sense of optimism might normalize his blood pressure. Similarly, a patient who uses homeopathy to deal with hay fever will expect the remedy to be helpful, hence the placebo effect may actually reduce the hayfever symptoms, or perhaps the patient tolerates the same symptoms with more fortitude – either way, the patient is happier. Some patients take homeopathy for self-limiting conditions, such as colds, which will clear up regardless after a week or so – in these sorts of cases, the placebo effect makes the patient feel better because he or she is given the illusion of taking control of the illness. For some conditions, such as back pain, conventional medicine struggles to offer a reasonably good solution, which means that a homeopathic remedy might be as good as anything else. After all, it will garner whatever psychological strengths the patient can bring to bear.

With all these undoubted benefits, it might seem that the use of homeopathy as a placebo is an obviously good thing, because it gives patients hope and relief. Many people might even argue that this is sufficient justification for homeopathy to be embraced by conventional doctors.

However, we take a different view. Despite the allure of the placebo effect, which is often (but not always) cheap, safe and helpful for patients, we strongly believe that it would be wrong for doctors and other healthcare practitioners to use homeopathic pills in this way. We base this stance on a variety of arguments.

One of our main reasons for discouraging the use of

placebo-based alternative medicine is the desire for honesty between doctor and patient. For the last few decades, the consensus in medicine has moved very decidedly towards encouraging a doctor–patient relationship based on openness and fully informed consent. This has involved doctors using the principles of evidence-based medicine to offer patients those treatments that hold out the greatest likelihood of success. Any reliance on placebo treatments would undermine all these goals.

Doctors who studied the research into, say, homeopathy would soon realize that it is bogus and that any benefit to the patient is due to the placebo effect. If a doctor nevertheless decided to prescribe homeopathy, then he or she would be forced to lie to the patient in order for the placebo to be effective. In short, the doctor would have to reinforce the patient's misplaced faith in the extraordinary power of homeo-pathy or perhaps even instil such a false belief. The question is simple – do we want our healthcare to consist of honest, evidence-based treatments or do we want it built upon a foundation of lies and deceit?

In fact, the best way to exploit the placebo effect is to lie excessively in order to make the treatment seem extra special. A doctor could use statements such as 'this remedy has been imported from Timbuktu', 'you're receiving the last supplies', 'the remedy has had a 100% success rate so far this year', 'this remedy neutralizes the most evil anti-matter in each of your cells'. Such statements will raise a patient's expectations, thereby increasing the likelihood and extent of the placebo response. In short, for maximum-strength homeopathy, the doctor would need to tell the biggest pork pies imaginable.

Doctors regularly exploited placebos in the past, as they had little else to offer patients, but modern medicine now has real treatments that have been tested and shown to be effective. We strongly feel that there should be no return to a medical system that relies on placebos – a view shared by the doctor and journalist Ben Goldacre:

> Whether mainstream medics would want to go back to the old ways and embrace the placebo-maximising wiles of the alternative therapists is an easy question: no thanks. The didactic, paternalistic, authoritative, mystifying mantle has passed to the alternative therapist, and to wear it requires one thing most doctors are uncomfortable with, dishonesty.

Our position – that the routine use of placebos is unacceptable because doctors should never lie to their patients – might seem draconian. Indeed, those who oppose our view would argue that the benefits of lying outweigh our ivory-tower ethical arguments. These opponents would feel that white lies are acceptable if they improve the health of patients. We would counter that routine peddling of placebos would lead to a widespread culture of deception in medicine, which would in turn result in a series of corrosive consequences for the medical profession. Imagine what healthcare would be like if doctors routinely prescribed placebo-based treatments, such as homeopathy:

1   Doctors would have to establish a conspiracy of silence and agree not to reveal the bogus nature of homeopathy. None of them would be allowed to point out the truth about the Emperor's New Clothes, as this would undermine homeopathy's placebo effect.

2   Medical researchers would not sign up to this pact, as their mission is to understand disease, its causes and its cures. In the name of progress, they would be honour bound to point out that existing research fails to support homeopathy. This would lead to scientists and doctors giving conflicting messages.

3   Homeopathic prescriptions would act as a gateway drug, encouraging patients to experiment with other irrational treatments. Professor David Colquhoun has neatly summarized the insidious dangers of homeopathic remedies: 'Their sugar pills contain nothing and they won't poison your body. The greater danger is that they poison your mind.'

4   Parents might ignore scientists who promote life-saving interventions such as vaccination, and instead they might listen to homeopaths promoting alternative (and ineffective) ways to protect children. After two centuries of progress since the beginnings of the Age of Enlightenment, a decision to move away from evidence-based medicine could usher in a New Age of Endarkenment.

5   Pharmaceutical companies would be in a strong position to argue that they too could promote placebo remedies. Why should they bother with the expensive process of proper drug development when they could make bigger profits by marketing a placebo sugar pill and pretending that it was a panacea?

Finally, there is one more reason why placebo treatments should be avoided. In fact, this particular reason is so powerful that it will soon become obvious that it is completely unnecessary and unjustifiable to use placebos routinely to treat patients. Everyone agrees that the placebo effect can be very beneficial, but the truth is that we do not need a placebo in order to evoke a placebo effect. Although this sounds paradoxical at first, it actually makes perfect sense if we explain what we mean in more detail.

Whenever a doctor prescribes a proven treatment, then the

patient hopefully experiences a biochemical and physiological benefit. However, it is important to remember that the impact of a proven treatment is always enhanced by the placebo effect. Not only will the treatment deliver a standard benefit, but it should also deliver an added benefit because the patient has an expectation that the treatment will be effective. In other words, patients receiving proven treatments already receive the placebo effect as a free bonus, so why on Earth would a patient take a placebo on its own which delivers only a placebo effect? And why on Earth would a therapist prescribe just a placebo? This would simply short-change the patient.

Doctors are well aware that all their treatments come with a placebo effect, the extent of which depends on a whole host of factors. These include the doctor's clothing, confidence and general attitude. The best doctors fully exploit the placebo impact, while the worst ones add only a minimal placebo enhancement to their treatments; this explains why the neurologist J. N. Blau suggested, 'The doctor who fails to have a placebo effect on his patients should become a pathologist.'

Earlier we outlined a list of conditions that homeopaths might treat which would improve due to the placebo effect. Returning to these same conditions, we can see that conventional doctors will generally advise a more reliable medical treatment that will not only offer a direct benefit to the patient, but will also offer an indirect benefit via the placebo effect. So, instead of recommending homeopathic Arnica for a severe bruise, a doctor might suggest a cold compress within the first day of the injury and then a damp, warm cloth thereafter. Instead of homeopathy for high blood pressure, a doctor might suggest a change in diet, or less alcohol consumption or fewer cigarettes, and if this does not work then the condition can also be treated with effective drugs. Similarly, for patients suffering from hay fever, a non-drowsy antihistamine that has been proved to work, plus its inevitable placebo effect, would be a much better option than a homeopathic placebo on its own. A cure for the common cold still eludes science, so conventional medicine can only treat this condition in terms of addressing the accompanying symptoms, but even this is more than homeopathy can achieve. The proven benefits of conventional cold tablets plus their placebo effect are, again, better than just the placebo effect of homeopathic tablets.

For the hardest problems, such as back pain, doctors have a limited arsenal of truly effective options, but these are still more powerful than anything that homeopathy or any placebo-based alternative therapy can offer. In 2006, B. W. Koes and his Dutch colleagues published a clinical review entitled 'Diagnosis and treatment of low back pain' in the *British Medical Journal*:

> The evidence that non-steroidal anti-inflammatory drugs relieve pain better than placebo is strong. Advice to stay active speeds up recovery and reduces chronic disability. Muscle relaxants relieve pain more than placebo, strong evidence also shows, but side effects such as drowsiness may occur. Conversely, strong evidence shows that bed rest and specific back exercises (strengthening, flexibility, stretching, flexion, and extension exercises) are not effective.

As we approach the end of this book, it becomes increasingly clear that much of alternative medicine is ineffective and should not be encouraged, even at the level of being a benign placebo. In many ways, today's alternative medicine is a modern version of the snake-oil remedies that were sold widely in America a century ago, such as Tex Bailey's Rattlesnake Oil and Monster Brand Snake Oil. They offered no medical benefit to patients, but they made plenty of profits for the huxters who sold them. One of the most famous snake-oil salesmen was Clark Stanley, who promoted his product as 'A Liniment that penetrates Muscle, Membrane and Tissue to the very bone itself, and banishing pain with a power that has astonished the Medical Profession'. Of course, it offered no such benefit, and when his Snake Oil Liniment was tested in 1916 it was found to be devoid of any actual snake oil. Instead, it consisted of 'principally a light mineral oil mixed with about 1 percent of fatty oil, probably beef fat, capsicum and possibly a trace of camphor and turpentine.'

Both snake oil and ultra-diluted homeopathic remedies contain no active ingredient, and both also offer nothing but a placebo effect. Yet the former is now mocked and seen only in Hollywood cowboy films, and the latter is still sold in every pharmacy. If anything, homeopathy is even more absurd than snake oil, as demonstrated by a homeopath who wrote a letter outlining a particularly bizarre homeopathic remedy: 'This patient continues to have multiple symptoms of lumps on scalp and has had a flu-like illness. Overall her mood has improved, however, I have given her a dose of Carcinosin Nosode 30C over the day followed by Berlin Wall 30C one a day in the morning . . .' A response in the *Medical Monitor* emphasized the ridiculous nature of Berlin Wall as a homeopathic remedy: 'What therapeutic advantages does Berlin Wall have over ordinary garden wall or Spaghetti Junction concrete? And do Scottish homeopaths use microdoses of that historic nostrum, Hadrian's Wall? I think we should be told.'

So how did we get into a position whereby each year we are spending £40 billion globally on alternative therapies, most of which are as senseless as homeopathy, and many of which are a good deal more dangerous? In the penultimate section of this book we will look at the ten groups of people who are most responsible for our increasing enthusiasm for alternative medicine. In each case, we will explain the role that the group has played in giving alternative medicine undue credibility, and moreover we will suggest how each of them can correct the prevailing overly optimistic, uncritical and misguided view of alternative medicine. What follows is an analysis of what has gone wrong over the last quarter of a century, coupled with a manifesto for re-establishing the role of evidence-based medicine.

Top 10 culprits in the promotion of unproven and disproven medicine

1 Celebrities

This list is compiled in no particular order, so celebrities are not necessarily the worst offenders in terms of the unjustified promotion of ineffective alternative medicine, but they have certainly played an important role in recent decades. When Professor Ernst and his colleague Max H. Pittler looked for articles published in 2005 and 2006 which involved well-known people using alternative medicine, they discovered several dozen famous figures who were being linked with various unproven therapies.

The celebrities ranged from fans of homeopathy, such as Pamela Anderson, Cindy Crawford and Cher, to devotees of Ayurvedic medicine, such as Goldie Hawn and Christy Turlington. These high-profile names give alternative medicine a greater level of credibility among the public, because they are clearly people who can afford the best medical treatment. In other words, these treatments may be perceived as superior to mainstream treatments because they are bought at a premium by the rich and famous.

In addition to actors and singers, there are also many sportsmen and sportswomen who have indulged in alternative medicine, such as Boris Becker and Martina Navratilova.

These sporting celebrities deliver extra credibility, because they are role models. We assume that they take special care of their health and have excellent advisers. The truth is that wealthy sportsmen and their coaches can afford to waste money on extravagant placebos, while also spending large sums on the very best that conventional medicine has to offer.

The US homeopath Dana Ullman clearly believes that celebrities help sell alternative therapy to the public, because his latest book, *The Homeopathic Revolution*, is subtitled *Why Famous People and Cultural Heroes Choose Homeopathy*. He tries to convince readers that homeopathy must work on the grounds that it has been used by some of the most famous figures in history, including eleven American presidents, seven Popes, Beethoven, Goethe and Tennyson, as well as Axl Rose, the lead singer of Guns N' Roses.

All these uninformed or ill-informed celebrities would do the public a service if they stopped endorsing useless therapies. Better still, celebrities should arm themselves with the best available evidence and condemn faddish, flawed and dangerous treatments. The singer Kylie Minogue did exactly this in 2005 when she issued a statement regarding rumours that she was using alternative therapies to treat her own cancer: 'She has asked her fans please not to believe stories of dramatic weight loss and desperate searches for alternative therapy. Kylie has made it clear to her representatives that she doesn't want fellow sufferers to be misguided by the false stories regarding her condition and her choice of doctors.'

Even more impressively, the actor Richard E. Grant exposed a dangerous scam involving goat serum as a life-saving treatment for HIV/AIDS. Having been brought up in Swaziland, Grant was invited to endorse goat serum in Africa, but his reaction was not the sort of thing that the vendors of goat serum had been looking for: 'Dead people are now Lazarused from the grave – Bullshit!' Grant acted responsibly and notified journalists working for BBC *Newsnight*, who publicized the fact that goat serum is just snake oil.

2 Medical researchers

This category may surprise many of our readers. After all, throughout this book we have relied upon medical researchers to investigate alternative medicine. It is only thanks to their efforts that it has become apparent that so many of these treatments are ineffective. Not only have they conducted research into alternative medicine, they have also done their best to disseminate the disappointing truth about the various therapies. However, most medical researchers do not investigate alternative medicine, but rather they focus on developing conventional treatments – our criticism is reserved for them.

There has been a general tendency for researchers to focus on their own speciality, perhaps developing new antibiotics, vaccines or surgical techniques, while ignoring the fact that alternative practitioners are often undermining their work by scaremongering about conventional medicine and overhyping their own alternative treatments. In other words, too many medical researchers have stood by and silently watched the rise of alternative medicine and the crackpot theories behind them.

There have been only a few shining examples of academics who have gone out of their way to highlight the contradictions, exaggerated claims and falsehoods within much of alternative medicine, but in many cases the consequences have been quite remarkable. In 2006, a loose coalition of like-minded scientists wrote an open letter to chief executives of the National Health Service Trusts, who are ultimately responsible for allocating funds for healthcare. The signatories, who included only one specialist researcher in the area of alternative medicine (Edzard Ernst), simply argued that homeopathy and many other alternative therapies were unproven and that the NHS should reserve its funds for treatments that had been shown to work: 'At a time when the NHS is under intense pressure, patients, the public and the NHS are best served by using the available funds for treatments that are based on solid evidence.'

The letter generated a front-page headline in *The Times*, which then led to radio and television coverage. For the first time, many members of the public were being informed about the false claims of homeopaths and were learning that their taxes were being wasted on bogus remedies. Moreover, the chief executives of the NHS Trusts seemed to pay some attention to the letter and reviewed their policies towards homeopathy – by mid-2007, 21 trusts continued their funding for homeopathy unabated and 40 trusts had not disclosed their spending on homeopathy, but 86 trusts had either stopped sending patients to the four homeopathic hospitals or were introducing measures to limit referrals strictly. Hilary Pickles, director of public health at Hillingdon Primary Care Trust, explained his view on funding homeopathy in *The Times*:

> It isn't just that there is no evidence base for homoeopathy; it is also a question of spending priorities. Every time you decide to spend NHS money on one thing, something else is losing out. It is completely inappropriate to spend money on homoeopathy that is unproven, as it means less money for other treatments that are known to be effective.

A group of vets embarked on a more satirical campaign against the use of homeopathy by forming the British Veterinary Voodoo Society in 2005. They were outraged when the Royal College of Veterinary Surgeons decided to publish a list of vets who practised homeopathy, which would effectively promote and tacitly endorse the practice. These anti-homeopathy vets, whose main concern was that animals should receive the best available treatments, were arguing that homeopathy was on a par with voodoo when it came to evidence and efficacy. Their campaigning has helped to persuade veterinary societies to behave more responsibly, and the Federation of Veterinarians in Europe (FVE) now urges its members 'to work only on the basis of scientifically proven and evidence-based methods and to stay away from non-evidence-based methods'.

Activism among medical researchers can be effective, so more need to stand up and be counted. A word of warning, however, because those who dare to question the value of alternative medicine can easily become the target of attacks on their reputation and integrity. They are often accused of being in the pay of the big corporations. One line of defence is to emphasise that researchers are generally driven by the desire to cure disease and increase the quality and length of human life.

For instance, Professor Michael Baum, who is a specialist in breast cancer and a signatory of the 2006 letter arguing against unproven treatments in the NHS, has adopted the following approach when lecturing on the subject of evidence-based medicine: 'I often introduce myself as a son of a mother, a husband of a wife, a brother of a sister, a father of two daughters and an uncle to seven nieces. My mother died tragically from breast cancer and my sister is a long-term survivor.' In other words, Professor Baum has both a personal and a professional interest in identifying the best treatments for breast cancer, and in particular the death of his own mother has inspired his dedication to saving lives.

3 Universities

Science degrees have always been a treasured commodity. Students who have successfully completed a Bachelor of

Science (BSc) degree have demonstrated that they have grasped the general principles and foundations of a particular discipline and are ready to study at a higher level. By earning a science degree, graduates have shown that they understand the knowledge derived from previous experiments and are close to conducting their own research. Or at least that is what science degrees used to represent. Today, some universities have decided to devalue the significance of the BSc by demeaning the traditions of science and prostituting the integrity of scholarship.

Universities around the world are now offering degrees in various forms of alternative medicine, which undermines everything that a university should stand for. How can a university offer a BSc degree course in alternative medicine teaching the principles of Ch'i, potentization and subluxations (key concepts in acupuncture, homeopathy and chiropractic therapy respectively), when they make no scientific sense whatsoever?

Such degrees do a disservice to students, who are given the false impression that they are learning the science behind a system of healthcare. At the same time, patients can also be misled, because they may hear that alternative medicine is being taught at university and will then assume that it must be effective. In short, universities give alternative medicine an undeserved level of credibility.

The completely crass nature of alternative-medicine degrees is easily demonstrated by a question posed in 2005 to students taking the 'Homeopathic Materia Medica 2A' examination at the University of Westminster, London: 'Psorinum and Sulphur are Psoric remedies. Discuss the ways in which the symptoms of these remedies reflect their miasmatic nature.' This question is a throwback to the Dark Ages of medicine, when it was believed that disease was caused by *miasmas,* which were poisonous vapours – this idea became obsolete in the late nineteenth century when scientists developed the more accurate and useful germ theory of disease.

Professor David Colquhoun surveyed the state of play in Britain in 2007 and discovered that there are sixty-one degree courses in alternative medicine, of which forty-five are BSc degrees, spread across sixteen universities. Five of these BSc degrees specialize in homeopathy – this means that students spend three years studying a subject that we have demolished in this book in a single chapter.

The worst offender seems to be the University of Westminster, which offers fourteen degrees in alternative medicine. This university offers many degrees in more respectable subjects and its staff in other departments have generally good reputations, so why has it started to offer meaningless degrees in phoney subjects? According to Colquhoun, the problem is that universities that offer courses in unproven therapies have prioritized profit above integrity:

> This is the equivalent of teaching witchdoctory. If you have a Bachelor of Science degree, it ought to be in something that can vaguely be described as science ... I'd like to see vice-chancellors get honest. They've lost their way and are happy to teach anything to get bums on seats. They think anything that makes money is OK. We know that these courses are showing bigger rises than any other subject, while maths and other subjects are going down.

It is time for those in responsible positions in universities to change priorities. Academic standards must not be sacrificed for financial considerations. A strategy that mainly aims at profit is shortsighted; it may be successful in the short term but in the long term it will undermine the integrity of our institutions of higher education.

4 Alternative gurus

It is strange that we live in an era when alternative practitioners are more famous than conventional practitioners. For example, the US health guru Deepak Chopra is a world-famous promoter of Ayurvedic medicine and other alternative therapies, and there is no conventional doctor who can match his global celebrity status.

Chopra and his fellow health gurus have been spreading the gospel of alternative medicine for well over a decade, achieving major press coverage, appearing on the most popular TV shows and lecturing to vast audiences. Their undeniable charisma, coupled with corporate professionalism, has meant that they have had a major impact on the public's perception of alternative medicine. In general, they have simply added to the often exaggerated and misleading claims surrounding these therapies.

For example, Dr Andrew Weil is one of America's most successful proponents of alternative medicine, having twice adorned the cover of *Time* magazine and regularly appearing on *Oprah* and *Larry King Live*. He labels himself 'Your trusted health advisor'. He does have a background in medicine, so some of his advice is sensible, such as encouraging more exercise and less smoking. However, much of his advice is nonsense, and the problem for his legions of followers is that they may not be able to tell the difference between the sensible advice and the nonsense. In *Natural Health, Natural Medicine*, published in 2004, he actively discourages readers from using prescription drugs for treating rheumatoid arthritis, even though some drugs can indisputably alter the course of the disease and offer the chance of preventing crippling deformities.

While sometimes denigrating conventional medicines that do work, Weil seems to encourage alternative therapies that do not work, such as homeopathy. He even suggests to patients that they should experiment with a range of alternative therapies and find out what works for them, which particularly concerned the retired physician Harriet Hall, who reviewed his book in *Skeptical Inquirer* magazine: 'The problem with this approach is many conditions are self-limited and others have variable courses. When your symptoms happen to subside, you will falsely attribute success to whatever remedy you happened to be trying at the time.' Rather than encouraging patients to self-experiment and come to possibly unreliable conclusions, it would be better if Weil accurately publicized the conclusions from carefully and safely conducted clinical trials.

Dr Weil's suck-it-and-see philosophy is shared by many of his fellow authors in the genre of alternative medicine. They readily throw every imaginable alternative treatment at their readers, as shown by Professor Ernst and his colleagues, who surveyed seven of the leading books on alternative medicine. Altogether, these books offered forty-seven different treatments for diabetes, of which only twelve appeared in more than one book. Five of these treatments (hypnotherapy, massage, meditation, relaxation and yoga) can help patients with their general wellbeing, but none of the other treatments for diabetes is backed by any evidence at all. There is a similar level of conflicting and misleading advice in relation to cancer – the seven books suggest a total of 133 different alternative treatments.

Kevin Trudeau is another high-profile guru – his book *Natural Cures They Don't Want You To Know About* has sold 5 million copies and topped the *New York Times* bestseller list. This success is baffling, as Trudeau has no medical credentials. Instead, his Wikipedia entry describes him as 'an American author, pocket billiards promoter (founder of the International Pool Tour), convicted felon, salesman, and alternative medicine advocate'. After serving two years in a federal prison for credit-card fraud, he worked in partnership with a company called Nutrition for Life. He soon fell foul of the law again, and was sued for essentially operating a pyramid-selling scheme. In his third incarnation, Trudeau started using TV infomercials to sell products, but over and over again he was accused of making false and unsubstantiated claims, so much so that in 2004 the Federal Trade Commission fined him $2 million and permanently banned him from 'appearing in, producing, or disseminating future infomercials that advertise any type of product, service, or program to the public'.

Although he can no longer promote products on TV, freedom of speech means that he can still appear on TV to tout his books, sometimes topping the table of infomercial appearances in a single week. His bestselling book contains such dangerous nonsense as 'The sun does not cause cancer. Sun block has been shown to cause cancer', and 'All over-the-counter non-prescription drugs and prescription drugs CAUSE illness and disease.' In 2005, the New York State Consumer Protection Board issued a warning that Trudeau's book 'does not contain the "natural cures" for cancer and other diseases that Trudeau is promising'. The Board also cautioned the public that 'Trudeau is not only misrepresenting the contents of his self-published book, he is also using false endorsements to encourage consumers to buy the book.'

Unfortunately, Trudeau seems like an unstoppable force, and he continues to sell alternative health products via his website. The New York journalist Christopher Dreher sees a clear strategy in Trudeau's business ambitions: 'In essence, the infomercial sells the book, which sells the Web site – which nets Trudeau tons of money.' Alternative-health gurus nearly always promote health products from which they benefit financially, either directly or indirectly. Even the mild-mannered, avuncular

Dr Weil does not shy away from a corporate approach to his role as a health guru, as demonstrated by his brand of alternative therapies sold under the banner of 'Dr Weil Select'. On top of this, in 2003/4 he received $3.9 million in royalties, having signed a deal with Drugstore.com.

Similarly, American radio host and self-proclaimed health visionary Gary Null markets products through his own website. Part of his marketing policy is to trash conventional medicine in order to promote the alternative, but this leads to some particularly irresponsible and dangerous proclamations. In his book *AIDS: A Second Opinion*, Null states: 'AIDS of the 90's has become an iatrogenic disease brought on or made worse by immunosuppressive drugs.' In other words, Null argues that conventional medicine harms rather than helps HIV/AIDS patients. Peter Kurth, a journalist infected with HIV, reviewed Null's book and did not pull his punches:

> Null's blithe disregard of the evidence seems less blinkered than criminal . . . And when the late Michael Callen is quoted as if he were still alive, I nearly jumped out of my skin. (Callen, once famous as a long-term survivor of AIDS and adamantly opposed to the use of AZT, has been dead since 1993.)

Another health guru with a strange view on treating HIV is Patrick Holford, a British-based alternative nutritionist who is the author of twenty-four books, which have been translated into seventeen languages. In 2007, his latest book was accused of making dangerous claims about the treatment of HIV. When he was in South Africa, he even repeated his claims to the press: 'What I have said in the latest edition of my book, the *New Optimum Nutrition Bible* . . . is that "AZT, the first prescribable anti-HIV drug, is potentially harmful and proving less effective than vitamin C."'

Holford's views have angered so many scientists over the years that he has even inspired a website entitled Holford Watch (www.holfordwatch.info) which seeks to highlight and correct his errors. Nevertheless, the University of Teesside judged Holford to be of sufficiently high standing to appoint him as a visiting professor. This links back to the problems highlighted in two of the previous subsections. First, some universities are acting in a peculiarly shoddy manner when it comes to alternative medicine. And second, medical researchers who should be making a fuss are not standing up for academic standards at their institutions.

5 The media

Newspapers, radio and television are, of course, hugely influential in any debate. However, the desire to attract readers, listeners and viewers means that the media are under pressure to sensationalize. This sometimes means not letting the facts get in the way of a good story.

This was demonstrated by a survey of Canadian print media by the Department of Community Health Sciences at the University of Calgary. Three researchers scanned nine publications for articles that appeared between 1990 and 2005, looking for any that linked CAM (complementary and alternative medicine) to cancer treatment. They found 915 articles in total, of which 361 had CAM treatment for cancer as the primary focus of the article. The main results confirmed previous, similar investigations:

> CAM therapies were most often described in a positive fashion, and CAM use was most often (63%) described as a potential cure for cancer. The majority of articles did not present information on the risks, benefits, and costs of CAM

uses and few provided a recommendation to speak with a health care provider before use.

In short, the print media in Canada (and elsewhere) tend to present an overly positive and simplistic view of alternative medicine. The way that alternative medicine is presented in newspapers all too often flies in the face of the evidence.

Turning to television, it seems that daytime programmes are always happy to invite a misguided alternative therapist onto their sofas. *The Wright Stuff*, for instance, is a largely reputable morning show on Channel Five in the UK, but it can arguably mislead its viewers with its regular slots for alternative practitioners. Jayney Goddard, President of the Complementary Medical Association (CMA), appears frequently on the show, usually promoting homeopathy. Chapter 3 explained that homeopathy is nothing more than a placebo, but innocent viewers of *The Wright Stuff* are generally given the impression that it is a powerful form of medicine.

It is interesting to note that the CMA's website claims: 'Thousands of people have contacted the CMA and *The Wright Stuff* about some of the products mentioned on the show by Jayney Goddard in the last few weeks.' This is a conflict of interest, inasmuch as Goddard admits that she helped to formulate a brand of supplements that were promoted on the programme and which are sold on the CMΛ's website. Such conflicts of interest turn out to be the rule and not the exception. The programme producers probably feel that they are simply filling fifteen minutes of airtime with some harmless medical chat, but they are actually encouraging a market in unproven treatments. Moreover, *The Wright Stuff* is indirectly promoting some rather peculiar views, as Goddard is the author of *The Survivor's Guide to Bird Flu: The Complementary Medical Approach*, which claims to offer: 'Information about a specific remedy for the precise symptoms of H5N1.' There simply is no alternative cure for bird flu, and to say otherwise is irresponsible in the extreme.

Daytime TV has a particular penchant for the truly wacky end of the alternative spectrum, such as miracle healers who have superpowers. In North America, Adam Dreamhealer has been a popular miracle healer ever since a large black bird informed him of all the secrets of the universe. His massive media presence would be comical if it were not for the fact that large numbers of patients put their faith in the supposed healing skills of Mr Dreamhealer. According to his website: 'Adam uses energy healing in a unique way to merge the auras of all participants with healing intentions. Then he uses holographic views to energetically affect through intention those present.'

The European equivalent of Dreamhealer is Natasha Demkina, who claims to be able to diagnose disease thanks to her X-ray vision, which she has had since she was ten years old: 'I was at home with my mother and suddenly I had a vision. I could see inside my mother's body and I started telling her about the organs I could see. Now, I have to switch from my regular vision to what I call medical vision. For a fraction of a second, I see a colorful picture inside the person and then I start to analyze it.' However, in 2004 she underwent scientific testing and failed to prove that she had X-ray powers.

That same year Demkina appeared on a British daytime TV show called *This Morning*. She examined the show's medical expert, Dr Chris Steele, and saw problems with his gallbladder, kidney stones, liver and pancreas. As reported by Andrew Skolnickin in *Skeptical Inquirer*: 'The physician rushed off to have a battery of expensive and invasive clinical tests – which found nothing wrong with him. In addition to being exposed to unnecessary diagnostic radiation, he had a colonoscopy, which is not without risks.' Studies show that

1 in 500 patients who undergo colonoscopic screening suffer a bowel perforation. Viewers who saw Demkina's appearance on the show were probably impressed by her proclamations. Even though a later programme revealed that her diagnosis turned out to be alarmist and potentially dangerous, only a fraction of the original viewers would have learned of Demkina's failure.

It might come as no surprise that daytime TV, tabloid newspapers and mass-market magazines are featuring bogus therapies and miracle healers, but it is disappointing when the world's most respected broadcasters stoop to similarly low standards. In Chapter 2 we discussed how the BBC showed a misleading sequence that implied that acupuncture could act as a powerful anaesthetic for open-heart surgery, which was part of a supposedly authoritative documentary on the evidence for acupuncture. The BBC has a much deserved reputation for high-quality television, but sometimes it seems to lose its critical faculties when it comes to alternative medicine.

For example, a BBC news item in 2005 featured a *bioresonance machine* that could supposedly cure smoking addiction, but this was nothing more than fake gadgetry. John Agapiou, a neurophysiologist at University College, London, complained to the BBC:

> The item presented a treatment where the 'wave pattern of nicotine' is allegedly recorded and then inverted, nullifying the effect of nicotine on the body . . . In short the entire piece was a credulous and uncritical advertisement for this treatment . . . Bioresonance does not work. There is no experimental or theoretical validity to this nonsense. No scientific knowledge is necessary to realize this, just a little critical thought or even a little googling... It was stated in the program that bio-resonance can be used to treat illness. In fact, proponents claim

that it is an effective treatment for cancer! It is not. I'm sure you can see that an uncritical report such as this conspires to put vulnerable people and their money into the hands of charlatans and is culpable in any damage to their health caused by delaying or even preventing their access to effective medical care.

Another good example of bad broadcasting, cited by Dan Hurley in *Natural Causes*, comes from CBS in America. Their flagship investigative news programme *60 Minutes* essentially created an entire market for one of the most dubious alternative treatments in recent years. In 1993 the programme ran a segment entitled 'Sharks Don't Get Cancer', based on the contents of a book with the same name. Written by a Florida businessman called Bill Lane, the book argued that shark cartilage could be used to treat tumours. Lane's evidence to back this treatment came from some very preliminary research and the observation that sharks rarely get cancer. In fact, the 'Tumor Register of Lower Animals' records that forty-two varieties of cancer (including forms of cartilage cancer) have been found in sharks and related species.

There was already a small market in shark cartilage as a cancer treatment, but the hype generated by *60 Minutes* triggered a rush for shark-based remedies. According to Lane, there were thirty new shark-cartilage products on the shelves within two weeks of the broadcast and within two years these products were generating $30 million per year.

Yet the preliminary research did not demonstrate with any confidence that shark cartilage was effective in treating cancer. Had this been a conventional pharmaceutical, then it would have been forced to undergo years of research in order to prove that it was safe and effective, and only then would it have been available through prescription. But, because this was a natural alternative product, then no such regulation and testing were deemed necessary. Instead, shark cartilage was being distributed to health stores across America, and cancer patients were clamouring for it.

Incidentally, this took a terrible toll on the shark population. For example, Holland & Barrett, the UK's largest chain of health-food shops, has admitted that it sourced its shark cartilage from Spiny Dogfish and the Blue Shark, both of which are classified as 'vulnerable species', which means that they carry a high risk of becoming extinct. In a letter to the Shark Trust, the company stated: 'Holland & Barrett will continue to sell shark cartilage due to customer demand, until such time that the species is classed as an endangered species.' The classification 'endangered' means a very high risk of extinction, as opposed to merely a high risk.

In the late 1990s, concerned that the public was being duped, scientists began submitting shark cartilage to the sort of rigorous clinical trials that it should have undergone before being widely promoted. One by one, the trials concluded that shark cartilage had no medical value. Today, we can see how *60 Minutes* had nationally promoted a treatment that in reality offered no benefit, causing thousands of people to waste millions of dollars.

Worse still, it seems that some cancer patients suffered as a direct result of being swept up in the fad. In 1997 the *New England Journal of Medicine* reported the case of a nine-year-old Canadian girl who had undergone surgery to remove a brain tumour. Doctors had recommended radiation and chemotherapy as a follow-up treatment, which would have given the girl a 50/50 chance of survival. Her parents, however, had been impressed by the publicity surrounding shark cartilage and decided to forgo the conventional treatment in favour of the alternative. This decision, according to the doctors, removed any chance of survival: 'Four months later, marked tumor progression was documented, and the patient subsequently died ... We find it difficult to understand how conventional treatments for childhood cancer can be repudiated in favour of alternative approaches for which any evidence of efficacy is lacking.'

6 The media (again)

Mass media is a powerful force for influencing the public, which is why it deserves two slots in this list of top ten culprits. In the previous section we explained that the media exaggerates the benefits of alternative medicine, but in this section we will focus on how newspapers and television also sensationalize the risks of conventional medicine.

A 1999 survey of British newspapers by Professor Edzard Ernst sampled four broadsheet newspapers on eight separate days and discovered 176 articles relating to medicine. Twenty-six of the articles concerned alternative medicine, and they were unanimously positive – it seems that alternative medicine is almost beyond criticism. By contrast, the remaining articles about mainstream medicine were roughly 60 per cent critical or negative.

Without doubt, certain aspects of mainstream medicine deserve to be criticized, but the problem here is that newspapers and broadcasters are trigger happy. They cannot resist turning minor issues into major scares or presenting tentative findings as serious threats to the nation's health. For example, there have been numerous scare stories over the years suggesting that mercury-based dental fillings are toxic. These include a 1994 news report entitled 'The Poison in Your Mouth', which was part of the BBC current-affairs series *Panorama*. There was, however, no real evidence to warrant these concerns.

In fact, a major study in 2006 confirmed numerous previous

investigations showing that the fears over mercury fillings were groundless. Researchers monitored the health of 1,000 children who had received either mercury fillings or mercury-free fillings. Over the course of several years there was no significant difference between the two groups in terms of their kidney function, memory, coordination, IQ and other qualities. Although this was the most important paper ever published in this field, the journalist and clinician Ben Goldacre made a very telling observation:

> As far as I am aware there is no *Panorama* documentary in the pipeline covering the startling new research data suggesting that mercury fillings may not be harmful after all. In the UK there is not a single newspaper article to be found. Not a word on this massive landmark study, published in the prestigious *Journal of the American Medical Association*.

In this particular case, the media merely scared the public away from mercury fillings and towards more expensive, less reliable options, which then require more visits to the dentist. In other episodes of media hysteria, the consequences are far more serious. For example, the news stories concerning the measles, mumps and rubella (MMR) triple vaccine have genuinely endangered the health of thousands of children. Reports have tended to exaggerate the significance of preliminary or insubstantial research that questions the safety of MMR, while ignoring the high-quality research that demonstrates that the MMR triple jab is the safest option for children.

Irresponsible media coverage has caused a significant drop in the number of parents vaccinating their children, which in turn has already led to several measles outbreaks – the threat of a substantial epidemic still looms. Perhaps the media takes such a cavalier attitude because it has forgotten the damage caused by measles. While measles is merely an inconvenience for most families, it will cause ear infections for 1 in 20 children, respiratory problems for 1 in 25, convulsions for 1 in 200, meningitis or encephalitis for 1 in 1,000, and death for 1 in 5,000 children. In 2006 a British child died after contracting measles, the first such death in the UK for fourteen years.

In effect, poor reporting has started to undo the work of generations of researchers, who have devoted their careers to the battle against disease. Maurice Hilleman, for example, was born into a poor Montana family in 1919, living on a single meal a day and sleeping in a bunk riddled with bedbugs. He witnessed how childhood diseases had decimated his community, which later inspired him to develop eight of the fourteen vaccines routinely given to children, including MMR. He lived just long enough to witness the controversy over his life-saving vaccine. His colleague Adel Mahmoud still recalls Hilleman's reaction:

It saddened him to see that knowledge was twisted in such a way to play into the hands of the anti-vaccine movement and not really appreciate what vaccines are all about. They are about protection of the individual, but also protection of the society so that you achieve herd immunity. Maurice believed in that and it pained him a lot to see what was happening in the UK.

The mass media must decide whether it wants to report medical issues responsibly in order to inform the public, or to report it luridly in order to create shocking headlines. Unfortunately, the media has a profit motive and a lack of discipline, so the latter option will probably continue to be too tempting, particularly in light of how easy it is to scaremonger. This was demonstrated by an article entitled 'Mysterious Killer Chemical', published in 2005, which highlighted the dangers of the chemical DiHydrogen MonOxide, sometimes called DHMO.

It's found in many different cancers, but there's no proven causal link between its presence and the cancers in which it lurks – so far. The figures are astonishing – DHMO has been found in over 95% of all fatal cervical cancers, and in over 85% of all cancers collected from terminal cancer patients. Despite this, it is still used as an industrial solvent and coolant, as a fire retardant and suppressant, in the manufacture of biological and chemical weapons, in nuclear power plants – and surprisingly, by elite athletes in some endurance sports. However, the athletes later find that withdrawal from DHMO can be difficult, and sometimes, fatal. Medically, it is almost always involved in diseases that have sweating, vomiting and diarrhoea as their symptoms. One reason that DHMO can be so dangerous is its chameleon-like ability to not only blend in with the background, but also to change its state. As a solid, it causes severe tissue burns, while in its hot gaseous state, it kills hundreds of people each year. Thousands more die each year by breathing in small quantities of liquid DHMO into their lungs.

In fact, DHMO is just a highfalutin name for plain water (H_2O), and the article was written by the Australian science journalist Karl Kruszelnicki to show how easy it is to scare the public. He went on to point out: 'You can give people this totally accurate (but emotionally laden, and sensationalist) information about water. When you then survey these people, about three-quarters of them will willingly sign a petition to ban it.'

7 Doctors

Doctors ought to be ambassadors for evidence-based medicine, combining the best information from research with their own experience and knowledge of the particular patient in order to offer the best treatment options. This should mean that they discourage alternative treatments which generally come under the headings of unproven, disproven, dangerous or expensive.

Regrettably, too many GPs seem to take an entirely different stance. The numbers vary from country to country, but a reasonable ballpark figure is that roughly half of GPs refer patients to alternative therapists, and many more will respond positively to the idea of their patients trying remedies from the alternative-health section of the local pharmacy or health-food store. This raises the question, why are so many GPs tolerating, promoting, or even using bogus treatments?

One explanation could be ignorance. Many doctors may not be aware that most homeopathic remedies contain absolutely no trace of any active ingredient. They may not realize that the latest trials for acupuncture indicate that it offers negligible or no pain relief beyond placebo. They may be oblivious to the risks associated with spinal manipulation, and uninformed about the highly variable evidence relating to herbal remedies. Therefore, doctors may be giving the benefit of the doubt to treatments that really ought to be avoided.

Another, perhaps more important, factor is that doctors are constantly dealing with patients who have coughs, colds, backaches and other conditions which are either difficult or impossible to treat. Many of these troubling ailments will disappear over the course of a few days or weeks, so doctors might advise plenty of rest, a day off work, some paracetamol pills, or simply carrying on as normal. Some patients, however, are disappointed by these sorts of suggestions, and they may pester the doctor for something more obviously medical. Hence, it might be expedient for doctors to recommend something that placates the patient and which might also help them deal with the symptoms via the placebo effect. This might mean encouraging a patient to try a herbal or homeopathic remedy from a health-food store or pharmacy, even though the doctor might be aware that there is no evidence to support the use of either option.

This approach to patients – fobbing them off with placebos – was touched upon earlier in the chapter. It is paternalistic and inevitably involves deception. It also has negative consequences, such as medicalizing minor conditions which should be simply left alone, endorsing bogus remedies and encouraging patients in the direction of acupuncturists, homeopaths, chiropractors and herbalists.

And introducing patients to alternative therapists in relation to a minor condition could act as a gateway to a longer-term reliance on alternative practitioners. In turn, this might lead to treatments that are both ineffective and expensive, and possibly even dangerous. Moreover, there is a likelihood that the alternative practitioners will go on to counsel against proven conventional interventions, such as vaccinations, or meddle with prescription drugs. This undermines the role of doctors and endangers patients' health.

One solution is for doctors to be more honest with patients ('In a few days you'll be fine.') Another solution, which is something of an awkward compromise, is to offer patients a so-called *impure placebo*, which is more ethical than a *pure placebo*. Homeopathy is a good example of a pure placebo, as its only impact is via the placebo effect, and there is no justification at all for using it based on any scientific evidence. By contrast, magnesium in the treatment of anxiety is a good example of an impure placebo. This is because magnesium cannot really treat straightforward anxiety, but it can successfully treat some very rare conditions that have symptoms similar to anxiety. Hence, a doctor who treats a patient complaining of anxiety with magnesium might conceivably be giving the perfect remedy, because the patient might have one of these rare conditions. In reality, however, it is much more likely that the magnesium will only alleviate the patient's anxiety through the placebo effect. This form of impure placebo is much more acceptable than a pure placebo, because we are avoiding complete lies. On the other hand, we are still dealing in half-truths, as opposed to complete truths.

So far we have two categories of problematic physician. First, there is the ignorant doctor who advises alternative medicine, but who is unaware that it does not really work. Second, there is the lazy doctor, who advises alternative medicine in order to satisfy patients with otherwise untreatable conditions. Both types actively steer some patients towards alternative medicine; but there is a third category – the inconsiderate doctor – who inadvertently frustrates patients so that they seek out alternative therapies.

Surveys from across the world show that users of alternative medicine are motivated at least in part by their disappointment with conventional medicine. Doctors may well do a good job getting the diagnosis and the treatment right, but many patients feel that other, equally crucial, qualities of 'good doctoring' are missing. They feel that their doctor has too little time, sympathy and empathy for them, whereas survey data confirm that patients consulting an alternative practitioner are particularly keen on the time and understanding they often receive. In a way, it seems as though some doctors delegate empathy to alternative practitioners.

We believe that there is an important message here: alternative medicine is not so much about the treatments we discuss in this book, but about the therapeutic relationship. Many alternative practitioners develop an excellent relationship with their patients, which helps to maximize the placebo effect of an otherwise useless treatment.

The message for mainstream medicine is clear: doctors need to spend more time with patients in order to develop better doctor–patient relationships. The average consultation is as little as seven minutes in some countries, and even the most generous countries struggle to achieve an average of fifteen minutes. Of course, increasing consultation times is easier said than done. Alternative therapists happily devote half an hour to each patient, because they are generally charging a great deal of money for their time. Extending consultation times with GPs would require greater government investment.

Finally, it is also worth mentioning a much rarer, but more serious problem. There are a few doctors who are genuinely convinced of the power of alternative medicine, despite the lack of evidence. In the most extreme cases, they will apply unproven treatments in the most inappropriate cases, thereby jeopardizing the health of patients. There are appalling instances of this from around the world, including the case of Sylvia Millecam, which was discussed earlier in this chapter. Three of the alternative therapists who treated Sylvia before her death had a formal medical background, so they were brought in front of the Amsterdam Medical Disciplinary Tribunal after her death. This resulted in one of them being struck off, while the other two were suspended. Similarly, in 2006, the UK's General Medical Council considered the behaviour of Dr Marisa Viegas, who had become a practising homeopath with her own private clinic. Dr Viegas had advised a patient to replace her heart medication with homeopathic remedies, and a short time later the patient died. The General Medical Council declared that she had died of 'acute heart failure due to treatment discontinuation', and therefore they suspended Dr Viegas.

8 Alternative medicine societies

A plethora of societies around the world claim to represent practitioners of various alternative therapies. In the UK alone there are about one hundred! They could be a huge force for good, helping to establish high standards, promoting good practice and ensuring ethical principles. They could also encourage the further testing of alternative therapies for both efficacy and safety. In particular, these societies should be clarifying what their practitioners can treat and confirming which conditions are beyond their abilities. Instead, far too many of them make unsubstantiated claims for their particular therapy and allow their practitioners to conduct all manner of inappropriate interventions.

All of these problems exist, for example, among the societies who represent chiropractors around the world. The chiropractic societies have so far failed to establish systems for recording the adverse effects of spinal manipulation, which would at least help to gauge accurately the hazards associated with chiropractic therapy. Moreover, as mentioned in Chapter 4, a UK survey shows that an intolerably high percentage of chiropractors violate the essential ethical and legal principle of informed consent. Yet the General Chiropractic Council does not seem to take action. And the General Chiropractic Council continues to promote chiropractic therapy for various inappropriate conditions, despite the lack of any evidence. Its website claims that chiropractic care can offer 'an improvement in some types of asthma, headaches, including migraine, and infant colic' – this is simply not true.

The American Academy of Medical Acupuncture makes even more exaggerated claims, citing a long list of medical conditions that 'have been found to respond effectively to Medical Acupuncture', which includes insomnia, anorexia, allergic sinusitis, persistent hiccups, constipation, diarrhoea, urinary incontinence, flatulence and severe hyperthermia. Again, of course, there is no significant evidence to support the use of acupuncture for any of these conditions.

It is also worth noting that many of these societies have generally been weak (possibly negligent) in terms of exposing bad practice. Worse still, when the Society of Homeopaths, based in Britain, was criticized for not taking a firm stand against the inappropriate use of homeopathy, it decided to suppress criticism rather than to address the central issue. Andy Lewis, who runs a sceptical and satirical website (www.quackometer.net), had written about the Society and the issue of homeopathic malaria treatments, which resulted in the Society asking the company that hosts his website to remove the offending page. In our opinion, the Society needs to improve in three ways. First, it ought to police its practitioners more thoroughly. Second, it ought to act publicly and promptly when serious complaints are made. Third, it should listen to its critics rather than silence them.

The community of scientists, on the other hand, encourages criticism and debate within its own ranks. For example, in 2007 the Cochrane Collaboration established the Bill Silverman Prize 'to acknowledge explicitly the value of criticism of The Cochrane Collaboration, with a view to help-ing to improve its work, and thus achieve its aim of helping people make well-informed decisions about health care by providing the best possible evidence on the effects of health-care interventions'. In stark contrast to the community of alternative therapists, here is an organization offering a prize for those who criticize its work. Bill Silverman, you may remember, was the paediatrician who questioned his own theories on caring for premature babies and indeed proved himself wrong.

As well as, in our opinion, inadequately policing its own ranks, it seems that the Society of Homeopaths encourages bad practice. It appears to promote misleading, inaccurate and potentially dangerous ideas. In 2007, on World AIDS Day, the Society organized an HIV/AIDS Symposium in London. A spokeswoman for the Society claimed that the conference was about alleviating the symptoms of AIDS. In fact, there is not a shred of evidence to suggest that homeopathy can ease AIDS symptoms. Worse still, the conference discussed far more ambitious claims. The speakers were Hilary Faircloch, a homeopath who already works with HIV patients in Botswana; Jonathan Stallick, author of a book entitled *AIDS: the homeopathic challenge*; and Harry van der Zee, who believes that 'the AIDS epidemic can be called to a halt, and homeopaths are the ones to do it'. The last thing that HIV/AIDS sufferers need is false hope and barmy remedies.

9 Governments and regulators

In his book *Bad Medicine*, the historian David Wootton writes, 'For 2,400 years patients believed that doctors were doing them good; for 2,300 years they were wrong.' In other words, for most of our history, most medical treatments have failed to treat most of our diseases effectively. In fact, most of the doctors from previous centuries harmed rather than healed our ancestors.

The turning point came with the arrival of scientific thinking, the clinical trial and government regulation to protect vulnerable patients from harm – both physical and financial. The snake-oil salesmen were gradually driven out of business and mainstream medicine was forced to show that its treatments were both safe and effective before they could be employed.

In some instances, it required tragic events in order to bring about regulation. Or, as Michael R. Harris, historian of pharmacy at the Smithsonian Institution, put it, 'The story of drug regulation is built of tombstones.' For example, in 1937, a Tennessee-based pharmaceutical company called S. E. Massengill Co. used diethylene glycol as a solvent in the production of a new antibiotic called Elixir Sulfanilamide. There were no regulations requiring pre-market safety testing, so the company only became aware that the solvent was toxic when patients began to report serious side-effects. Typically, children were taking the elixir for a throat infection and were then suffering kidney failure and going into convulsions. The error caused over 100 fatalities, including the death of the company's chemist, Harold Watkins, who committed suicide when the scandal emerged. The following year American legislators passed the Federal Food, Drug, and Cosmetic Act, which allowed the Food and Drug Administration (FDA) to demand proof that new drugs were safe before going on sale. Regulations were still inadequate in many other parts of the world, but the Thalidomide tragedy of the 1960s compelled many other governments to bring in legislation. The UK Medicines Act of 1968, for instance, was a direct consequence of the Thalidomide disaster.

Alternative medicine, however, seems to have sidestepped these regulations. Buzzwords such as 'natural' and 'traditional' have allowed them to carry on largely unhindered in a parallel universe that is oblivious to safety issues. For example, in most countries, herbal remedies and other supplements can be marketed without rigorous proof of safety. The burden of proof is reversed: it is not the manufacturer who has to demonstrate that his product is harmless, but it is the regulator who has to prove that the product is harmful – only then can it be withdrawn from the market. This obviously is haphazard, as there are far too many products, so regulators react only when problems emerge. This is much like drug regulation before Thalidomide: a disaster (or several) waiting to happen.

Similarly, alternative practitioners tend to be un- or under-regulated. There are, of course, considerable national differences, but in general alternative practitioners do not require any in-depth medical training or experience. Indeed, literally anyone reading this text in Britain could call themselves a homeopath, a naturopath, a herbalist, an aroma-therapist, an acupuncturist, a reflexologist or an iridologist. You might have no training in conventional or alternative medicine, yet nobody could stop you nailing a sign to your front door and placing an advertisement in your local newspaper. It goes without saying that this situation is less than satisfactory. Serious diagnoses can be missed, conditions that never existed can be diagnosed, ineffective or harmful treatments can be applied, wrong or dangerous advice can be issued, and patients can be ripped off – and all this without adequate control or recourse.

By taking such a relaxed attitude towards alternative medicine, governments have exposed the public to medicines that are often ineffective and occasionally dangerous, and they have allowed alternative therapists, often deluded and occasionally disreputable, to ply their trade without hindrance. It would seem obvious that governments ought to be playing a more active role, by banning dangerous or useless alternative therapies and properly regulating those that are harmless and beneficial. Yet most governments have shied away from taking such a stance. For some reason they seem frightened of confronting the multi-billion-dollar alternative medicine industry. Or perhaps they are more worried about the millions of voters who currently use alternative medicine and who might be offended if their favourite herbalist or homeopath were forced to shut up shop.

There are numerous examples that demonstrate the need for governments to intervene, either by banning certain products or by tightly regulating them. For example, it is still possible to buy homeopathic kits for malaria protection on the internet or in your local health-food shop. One product claims to be 'a credible and highly effective alternative to conventional malaria treatments . . . Taken daily as a spray under the tongue it is suitable for all from the toughest adult to the tiniest tot: it even tastes good.' At a price of just £32.50 it seems like a bargain, except it does not work! Nobody seems to be enforcing any advertising or trading standards, and nobody seems to be worried about the public-health issue that is at stake here.

Governments ought to be moving rapidly to regulate therapists and products in order to protect patients, but there are very few signs that this will happen anytime soon. Indeed, there are clear signs that the British authorities are moving in the opposite direction, as they seem keen to encourage the use of largely unproven treatments. Two examples serve to demonstrate the desire of UK officials to return to the Dark Ages.

First, the UK Department of Health helped to fund a 56-page booklet written by the Prince of Wales's Foundation for Integrated Health. Entitled *Complementary Health Care: A Guide for Patients*, this has been one of the most influential documents in relation to alternative medicine, because it purports to be a reliable source of information for patients, and it was also distributed to every British GP. However, the booklet implies that alternative medicine is effective for a whole range of conditions, when we know that this is simply not the case, or at the very least we know that the evidence is poor.

For instance, the booklet states: 'Homeopathy is most often used to treat chronic conditions such as asthma; eczema; arthritis; fatigue disorders like ME; headache and migraine; menstrual and menopausal problems; irritable bowel syndrome; Crohn's disease; allergies; repeated ear, nose, throat and chest infections or urine infections; depression and anxiety.' Notice that the booklet does not say that homeopathy is effective for these conditions, but the phrase 'most often used to treat' certainly implies that patients should consider using homeopathy in all these situations. This government-subsidized propaganda is similarly misleading for chiropractic, herbal medicine, acupuncture and other forms of alternative medicine.

The Department of Health tried to defend the booklet's lack of rigour by declaring that it was never intended to include any scientific evidence about effectiveness, but this was less than honest. Professor Ernst had originally been asked to contribute to a whole section about scientific evidence, but this part was discarded before publication as such information presumably would have undermined the booklet's ambitions. Also, correspondence between the Department of Health and the Foundation for Integrated Health (obtained by Les Rose under the Freedom of Information Act) clearly shows that the guide was originally meant to include reliable information on effectiveness. In any case, if a patient guide does not contain such information, what on Earth is it for?

In a second example of British disinformation, the Medicines and Healthcare Products Regulatory Agency (MHRA) in 2006 took the shocking decision to allow homeopathic products to make claims on their labels based on homeopathy's own theory of testing known as 'provings'. As discussed in Chapter 3, these tests cannot demonstrate clinical effectiveness, and yet customers will now encounter labels based on provings and endorsed by the MHRA. This will mislead consumers into believing that homeopathic products are effective. The MHRA, which is an executive agency of the Department of Health, makes the proud claim: 'We enhance and safeguard the health of the public by ensuring that medicines and medical devices work and are acceptably safe.' Yet, for the first time since the creation of the Medicines Act, they have sacrificed their integrity.

The reason for the MHRA's shameful irresponsibility is hard to fathom, but Professor David Colquhoun feels strongly that the Prince of Wales has been an influential figure in this regrettable endorsement:

The MHRA have received letters from the Prince of Wales, and we are aware that an MHRA member has met the Prince at

Clarence House at least once. But all the contents are secret from the public. The Chairman of the MHRA Agency Board, Prof Alasdair Breckenridge, and chairman of their Herbal Medicines committee, Prof Philip Routledge, have both admitted to receiving such letters from the Prince of Wales, but neither will give any details, despite having been condemned by their own professional organisation, the British Pharmacological Society.

The MHRA argue that it is better to regulate and allow homeopathic remedies for safety reasons, but even if this were a good idea (and we do not think it necessarily is), then there still would be no need to give misleading indications of efficacy. Professor Michael Baum commented, 'This is like licensing a witches' brew as a medicine so long as the bat wings are sterile.' Journalist and broadcaster Nick Ross was equally scathing: 'Sometimes politics must take priority over science. After all, Galileo capitulated to the Inquisition. But what instruments of torture threatened members of the MHRA – or were they simply intellectual cowards?'

10 World Health Organization

This list of people, organizations and entities responsible for the unwarranted growth of ineffective and sometimes dangerous alternative medicine has been in no particular order, except that the World Health Organization (WHO) has been deliberately chosen to complete the list as it holds a special position.

No organization has done more to improve health around the world, such as the eradication of smallpox, and yet the WHO has acted shamefully in its attitude and actions towards alternative medicine. We would have expected it to provide clear and accurate guidance about the value of each popular alternative therapy, yet (as we saw in Chapter 2) in 2003 the WHO muddied the waters by publishing a highly misleading document on the value of acupuncture. Entitled *Acupuncture: Review and analysis of reports on controlled clinical trials*, the report based its conclusions on several unreliable clinical trials and thus endorsed acupuncture as a treatment for over 100 conditions. Of course, the evidence from high-quality reliable clinical trials paints a very different picture. In reality, acupuncture might possibly (though it looks less possible as each year passes) be effective in treating some types of pain and nausea, but it offers no proven benefit for any other conditions.

Naturally, ever since its publication, acupuncturists have cited the WHO report as the most authoritative evaluation of their mode of healing. And, not surprisingly, prospective patients have been persuaded that acupuncture must be effective for a whole range of conditions, because, after all, it has the blessing of the WHO. However, the WHO report was a shoddy piece of work that was never rigorously scrutinized and which should never have seen the light of day.

The WHO could repair its reputation if it were prepared to re-evaluate acupuncture fairly and publish a new report that reflected the evidence from the latest and most reliable trials. In this way, it could make a huge contribution to the public's understanding of what acupuncture can and, more often, cannot treat. Unfortunately, there is no sign that this is likely to happen.

Worse still, it seems that history is about to repeat itself and that the WHO is destined to fail us and embarrass itself again. According to a report in the *Lancet*, the WHO is planning to publish a report on homeopathy, which will have much in common with its irresponsible report on acupuncture. In other words, it will be rose-tinted and lacking in rigour.

Once again, practitioners will use the report to help validate invalid treatments. And, once again, patients will be persuaded that it is worth spending their money and risking their health on bogus treatments. For example, those who have seen a preliminary version of the report state that the WHO views homeopathy as a valid form of treatment for diarrhoea. Globally, over a million children die each year of diarrhoeal diseases, and an increased use of homeopathy would only make the situation worse. India's National Rural Health Mission is already showing signs of advocating homeopathy to treat diarrhoea, and the WHO report would only give credibility to this foolhardy policy.

The future of alternative medicine

The Scottish distiller Thomas Dewar once said: 'Minds are like parachutes. They only function when open.' On the other hand, the *New York Times* publisher Arthur Hays Sulzberger stated: 'I believe in an open mind, but not so open that your brains fall out.'

Of course, Dewar and Sulzberger both had a point, and their views were combined in a lecture in Pasadena in 1987, when the great American physicist Carl Sagan explained how science should treat new ideas:

It seems to me what is called for is an exquisite balance between two conflicting needs: the most skeptical scrutiny of all hypotheses that are served up to us and at the same time a great openness to new ideas. If you are only skeptical, then no new ideas make it through to you. You never learn anything new. You become a crotchety old person convinced that nonsense is ruling the world. (There is, of course, much data to support you.) On the other hand, if you are open to the point of gullibility and have not an ounce of skeptical sense in you, then you cannot distinguish useful ideas from worthless ones.

If all ideas have equal validity then you are lost, because then it seems to me, no ideas have any validity at all.

Throughout this book we have tried to strike a balance by being open to all forms of alternative medicine and all their respective claims, while submitting each one to the ordeal of testing. In general, the key test has been the clinical trial. Pioneered 250 years ago by James Lind and then refined over the course of the next century by Alexander Hamilton, Pierre Louis and many others, the clinical trial remains a beautifully simple, yet powerful, mechanism for getting to the truth. Indeed, Pierre Louis' description of a clinical trial still holds true today:

> For example, in any particular epidemic, let us suppose five hundred of the sick, taken indiscriminately, are subjected to one kind of treatment, and five hundred others, taken in the same manner, are treated in a different mode; if the mortality is greater among the first than among the second, must we not conclude that the treatment was less appropriate, or less efficacious in the first class than in the second?

Having sought to be both open-minded and sceptical, and having relied on all the best available evidence, our broad conclusion is fairly straightforward. Most forms of alternative medicine for most conditions remain either unproven or are demonstrably ineffective, and several alternative therapies put patients at risk of harm.

There will always be new research that will add to our knowledge, and it is possible that alternative treatments that currently appear ineffective might turn out to offer a significant benefit. However, while writing this book during the course of 2007, there have been major new studies that have only further undermined the credibility of alternative medicine. One of the most important was published in the *British Medical Journal* under the title 'Acupuncture as an adjunct to exercise based physiotherapy for osteoarthritis of the knee: randomised controlled trial'. The researchers gave advice and exercise to 352 patients, and then one-third received nothing else, one-third received real acupuncture, and one-third received sham acupuncture via the stage-dagger needles described in Chapter 2. The researchers concluded: 'Our trial failed to show that acupuncture is a useful adjunct to a course of individualised, exercise based physiotherapy for older adults with knee osteoarthritis.' This conclusion was reinforced by Eric Mannheimer's analysis of all the latest data, also published in 2007. These results were a serious blow for acupuncturists, who have argued that acupuncture for knee osteoarthritis was their most effective intervention. This particular treatment was even singled out for a special mention when the Prince of Wales addressed the WHO in 2006. It now seems that the jewel in the acupuncturists' crown is fake.

We wrote this book because we wanted to provide people with the most important research about alternative medicine, in the hope that readers would be in a better position to make informed decisions about their own healthcare. But what about those people who have not read this book? What about the millions of patients who have only been exposed to the media hype in the newspapers, exaggerated claims on the internet and misleading adverts in shop windows? Is it fair that they may waste their money and risk their health by using alternative medicine?

One of the greatest problems is that patients have virtually no protection when they enter the world of alternative medicine. Homeopathic remedies, for instance, are available on the internet, in high-street pharmacies and from anyone who claims to be a homeopath – in each case the vendors are selling treatments under false pretences, as homeopathic remedies are disproven and illogical. Similarly, couples seeking fertility treatment can waste large amounts of money on herbal treatments when there is no proper evidence or reason why they should be effective. Meanwhile, chiropractors expose their patients to large doses of X-ray radiation, manipulate the fragile bones of infants and apply heavy forces to adult necks, even though these treatments, in many cases, are totally ineffective. And so the story continues, with acupuncturists, reiki healers, psychic healers, shiatsu therapists and many other alternative practitioners making totally unfounded, yet hugely enticing, claims.

Notably, if any conventional doctor made such ludicrous promises and offered similarly unproven and even risky remedies, then he or she would be struck off or would perhaps end up in the dock.

Conventional medicine and alternative medicine both have the same ambition, namely to cure the sick, and yet one is tightly regulated and the other operates in the medical equivalent of the Wild West. This means that patients who venture towards alternative medicine are at risk of being exploited, losing their money and damaging their health.

The solution, surely, would be to create a level playing field, whereby alternative medicine has to maintain the same high standards required of conventional medicine. Regulation across the board would provide protection to all patients seeking any form of medical treatment.

In particular, this would mean that each alternative treatment would have to be tested, and only if it were proved that it generated more good than harm would it be permitted. Most patients are unaware of the immense amount of testing undergone by conventional treatments, so it is worth quickly summarizing how, for example, pharmaceuticals are assessed and investigated in order to see the sort of scrutiny that we are also proposing for alternative treatments such as herbal medicines.

Regulation and associated testing procedures vary across the world, but America is home to the biggest pharmaceutical industry, and its regulations are fairly typical of those found in many developed countries. The road from early-stage research to drugs available to patients can be broken down into six stages:

1   **Preclinical Research.** Scientists test different chemicals to see if they might have a role in medicine. This is likely to involve preliminary testing on animals to see if the chemical is likely to be sufficiently safe and effective. It will also take into consideration how the chemical might be mass produced if it turns out to be useful. This takes at least five years.

2   **Clinical Studies, Phase I.** The potential drug is given to between 10 and 100 volunteers to investigate safety in humans. The main goal of Phase I clinical studies is to identify a safe range of dosage. This takes between one and two years and costs roughly $10 million.

3   **Clinical Studies, Phase II.** The drug is given to between 50 and 500 patients with a relevant disease. The main goal is to gauge the effectiveness of the drug in humans. At the same time it is important to establish the optimum dosage and duration of treatment for the next stage of testing. Phase II takes two years and may cost a further $20 million.

4   **Clinical Studies, Phase III.** The drug is given to hundreds or thousands of patients to determine its effectiveness and any side-effects. This usually involves randomized clinical trials and the drug is tested against a control group receiving a placebo or the best existing drug. For thoroughness, Phase III may involve two

independent studies. It might generate a further stage of research if the drug seems to be effective only for a subset of patients, such as those in the early stages of the disease. Phase III takes three to four years and may cost $45 million.

5   **Review by Food and Drug Administration (FDA).** If Phase III has been successful, then it is possible that news of a breakthrough will reach the public. Before the drug is made available, however, the evidence has to be reviewed by the FDA in America, and its counterparts elsewhere, such as the European Medicines Evaluation Agency. This takes a further one or two years.

6   **Post-marketing Surveillance.** Even when the drug has passed all the tests and is being prescribed or sold over the counter, doctors will still be alert to any adverse reactions and report them to the FDA. This ongoing monitoring is important just in case there is a small risk that was not identified at Phase III.

These inordinate costs were cited in the journal *Scientific American* in 2000, so they have probably increased significantly in the last few years. Moreover, only a third of those drugs that enter Phase I trials reach the stage of New Drug Application, which allows a drug to be marketed to the public.

In short, conventional medicines have to leap over extraordinarily high hurdles, which requires a vast amount of money and time, but this level of testing is essential if the public is to be protected from harmful and ineffective drugs. Such rigorous testing should bring reassurance to patients, and it would not be unreasonable to demand similar levels of testing and regulation for alternative medicine.

There is, however, an argument that suggests that the sequence of testing for alternative medicines should be reversed. This is for the simple reason that these treatments are already widely available, whereas the pharmacological approach outlined above was developed for entirely new drugs. If millions of people are already using treatments such as herbal remedies and acupuncture, then it seems most sensible to assess the safety of these treatments based on the experiences of patients. For instance, practitioners could be asked to note the details of any adverse reactions and post them to a central database. The next priority would be to submit alternative therapies to clinical trials in order to find out for what (if any) conditions they are effective. Finally, if these tests were positive for a particular alternative therapy, scientists might investigate the mechanism behind the therapy and conduct the preclinical research.

This reversed approach puts patient safety first, because the consumer is already being exposed to alternative medicine, but it ultimately demands the same level of scientific rigour. Conducting such rigorous testing would be expensive, but remember that alternative medicine is a billion-dollar global industry, so it would not be unfair for it to devote part of its vast profits towards properly testing the products that it sells to the public. Moreover, governments already have funds for medical research, and they could act in a coordinated manner to devote a fraction of this money towards high quality clinical trials focused on the biggest selling products and the most popular therapies.

The reversed testing procedure, like the normal procedure, would take several years to complete. In the meantime, while awaiting the results, governments could require alternative remedies to carry labels and compel therapists to make disclosures that accurately reflect the existing evidence. Dylan Evans suggested exactly this idea in his book *Placebo*. For homeopathic remedies, he suggested the following label:

Homeopathy

Warning: this product is a placebo. It will work only if you believe in homeopathy, and only for certain conditions such as pain and depression. Even then, it is not likely to be as powerful as orthodox drugs. You may get fewer side-effects from this treatment than from a drug, but you will probably also get less benefit.

We believe that Evans's idea has some merit, because such open, honest and accurate summaries would certainly help patients. For alternative treatments supported by evidence, the summaries would contain the sort of helpful advice that we see on conventional pharmaceuticals. For other alternative treatments, the summaries would read more like a statutory health warning, akin to those found on cigarette packets.

As we approach the final few pages of the final chapter, it is interesting to see what happens if we apply Evans's idea to a selection of alternative therapies discussed elsewhere in this book. In some cases, the Evans-style summaries would appear on boxes of tablets, while in other cases they might appear on websites or in leaflets distributed in a clinic. And in every case, the summaries would help patients to be more aware of the evidence relating to any particular therapy.

Acupuncture

Warning: this treatment has shown only very limited evidence that it can treat some types of pain and nausea. If it is effective for these conditions, then its benefits appear to be short-lived and minor. It is more expensive than conventional treatments, and very likely to be less effective. It is likely that its major impact is as a placebo in treating pain and nausea. In the treatment of all other conditions, acupuncture either has no effect other than a placebo effect. It is a largely safe treatment when practised by a trained acupuncturist.

Chiropractic

Warning: this treatment carries the risk of stroke and death if spinal manipulation is applied to the neck. Elsewhere on the spine, chiropractic therapy is relatively safe. It has shown some evidence of benefit in the treatment of back pain, but conventional treatments are usually equally effective and much cheaper. In the treatment of all other conditions, chiropractic therapy is ineffective except that it might act as a placebo.

Herbal Medicine – Evening Primrose Oil

Warning: this product is a placebo. It will work only if you believe in it, and only for certain conditions which respond to placebo treatments. Even then, the placebo effect is unpredictable and it is not likely to be as powerful as orthodox drugs. You may get fewer adverse side-effects from this treatment than from a drug, but you will probably also receive less benefit.

Herbal Medicine – St John's Wort

Warning: this product can interact with other drugs – consult your GP before taking St John's wort. There is evidence that it is effective in the treatment of mild and moderate depression. Conventional drugs are available for these conditions, and are similarly effective.

These summaries reflect the broad range and complexity of alternative therapies, which includes treatments that are untested, or unproven, or disproven, or unsafe, or placebos, or only marginally beneficial, or almost certainly beneficial. Of all the above treatments, St John's wort has the most positive summary. Indeed, the clinical trials for St John's wort are so positive that GPs and scientists would endorse its use. Conventional medicine has no prejudice against any alternative treatment that can prove its worth, both in terms of safey and efficacy.

Fish oil is another excellent example of an alternative treatment that has been embraced by conventional medicine. Fish oil, available in capsules, comes under the heading of food supplements, and such supplements are discussed in the appendix. The trigger for detailed research into the possible benefits of fish oil was the observation that Inuits have very low rates of heart disease. This gave rise to further epidemiological investigations in other populations and eventually to clinical trials that have been uniformly positive. Ultimately, this has led to reassurances that fish oil is both safe and effective as a long-term preventative treatment for coronary heart disease. Detailed evaluation has also suggested that daily fish-oil capsules can extend life by one year on average. For those who do not eat oily fish on a regular basis, fish-oil capsules offer a clear benefit. Fish oil may also help control inflammation, which would be beneficial for people with arthritis or a range of skin problems.

Fish oil and St John's wort are marvellous examples of treatments that have emerged from traditional roots, which were then promoted within alternative medicine, and which have now been accepted by conventional medicine. Fish oil, in particular, is so utterly mainstream that it is no longer considered alternative by most conventional doctors, and St John's wort should be heading in the same direction. The appendix includes several other alternative therapies that conventional doctors would also endorse, particularly those that increase general wellbeing by relaxation or stress reduction – for instance, meditation and massage therapy.

This brings us to an interesting situation: any provably safe and effective alternative medicine is not really an alternative medicine at all, but rather it becomes a conventional medicine. Therefore, alternative medicine, by definition, seems to consist of treatments that are untested, or unproven, or disproven, or unsafe, or placebos, or only marginally beneficial.

Yet, alternative therapists continue to wear the name 'alternative' as a badge of honour, using it to give their substandard treatments an undeserved level of dignity. They use the term 'alternative' to promote the notion that they somehow exploit alternative aspects of science. The truth, however, is that there is no such thing as alternative science, just as there is no alternative biology, alternative anatomy, alternative testing, or alternative evidence.

Science, as we demonstrated in Chapter 1, is a universal approach for establishing the value of any medical intervention. The results of science are never complete and perfect, but step by step they bring us closer to the truth. The term 'alternative' is merely an attempt to escape from this truth by replacing the knowledge derived from science by hunches derived from other sources. This includes intuition, anecdote and tradition, which means that alternative medicine is based on personal opinions, the opinions of others and the opinions of our forefathers. However, in our introduction we pointed out:

> *'There are, in fact, two things, science and opinion;*
> *the former begets knowledge, the latter ignorance.'*

Even though Hippocrates wrote these words more than 2,000 years ago, it took us a phenomenally long time to really take this message seriously. When we finally did, about 150 years ago, medicine began to move rapidly out of the Dark Ages and doctors abandoned treatments such as bloodletting, which were more dangerous than the conditions they claimed to cure. Since then, progress has been immense and continuous. Immunization has eradicated killer infections; formerly fatal diseases affecting millions, like diabetes, appendicitis and many others, are now treatable; childhood mortality is only a fraction of what it once was, pain can be effectively controlled in most cases; and generally we live longer and enjoy a better quality of life. All of this is thanks to applying rational scientific thought to healthcare and medicine.

By contrast, the concept of an alternative type of medicine is a throwback to the Dark Ages. Too many alternative therapists remain uninterested in determining the safety and efficacy of their interventions. These practitioners also fail to see the importance of rigorous clinical trials in establishing proper evidence for or against their treatments. And where evidence already exists that treatments are ineffective or unsafe, alternative therapists will carry on regardless with their hands firmly over their ears.

Despite this disturbing situation, the market for alternative treatments is booming and the public is being misled over and over again, often by misguided therapists, sometimes by exploitative charlatans.

We argue that it is now time for the tricks to stop, and for the real treatments to take priority. In the name of honesty, progress and good healthcare, we call for scientific standards, evaluation and regulation to be applied to all types of medicine, so that patients can be confident that they are receiving treatments that demonstrably generate more good than harm.

If such standards are not applied to the alternative medicine sector, then homeopaths, acupuncturists, chiropractors, herbalists and other alternative therapists will continue to prey on the most desperate and vulnerable in society, raiding their wallets, offering false hope, and endangering their health.

Appendix:
Rapid Guide to
Alternative
Therapies

THE CORE OF OUR BOOK HAS FOCUSED ON ONLY FOUR OF THE main alternative therapies (acupuncture, homeopathy, chiropractic therapy and herbal medicine), but we have also evaluated many other therapies, which will be discussed in this section. We have devoted a page to each one, in which we address key issues, such as how did the therapy start, what does it involve, is it effective and is it safe? Despite the brevity of the sections, we have rigorously examined the scientific evidence for and against each therapy in order to reach our conclusions. You can also find more information about each alternative therapy in *The Desktop Guide to Complementary and Alternative Medicine: An Evidence-Based Approach*, a detailed reference book edited by Edzard Ernst, Max H. Pittler, Barbara Wider and Kate Boddy. This book also contains references to all the research that leads to the conclusions in this appendix.

Anybody considering alternative medicine as a mode of treatment, including all the therapies that follow, should take into account five pieces of advice. First, if you are thinking of using any form of alternative therapy for a particular condition, then we strongly recommend that you first consult and inform your GP – the treatment that you have chosen might interfere with any ongoing conventional therapies. Second, do not stop your conventional treatment unless your doctor advises that this is sensible. Third, bear in mind that alternative therapies can be expensive, particularly if they involve long-term consultations, so make sure that there is evidence to support the efficacy of a therapy before investing significant sums of money in its claimed benefits. Fourth, all therapies can generate placebo effects, but this alone is not enough to justify their use. Fifth, remember that every treatment carries risks, so make sure that the risks are outweighed by the benefits.

Included in this section are our evaluations of:

- Alexander Technique
- Alternative Diagnostic Techniques
- Alternative Diets
- Alternative Exercise Therapies
- Alternative Gadgets
- Anthroposophic Medicine
- Aromatherapy
- Ayurvedic Tradition
- Bach Flower Remedies
- Cellular Therapy
- Chelation Therapy
- Colonic Irrigation
- Craniosacral Therapy (or Cranial Osteopathy)
- Crystal Therapy
- Cupping
- Detox
- Ear Candles
- Feldenkrais Method
- Feng Shui
- Food Supplements
- Hypnotherapy
- Leech Therapy
- Magnet Therapy
- Massage Therapy
- Meditation

- Naturopathy
- Neural Therapy
- Orthomolecular Medicine
- Osteopathy
- Oxygen Therapy
- Reflexology
- Reiki
- Relaxation Therapies
- Shiatsu
- Spiritual Healing
- Traditional Chinese Medicine

Alexander Technique

Alexander technique is a therapy based on relearning correct postural balance and coordination of body movements. Fredrick M. Alexander was an Australian actor whose career was threatened by a recurring loss of voice. Doctors were unable to diagnose any problem with his throat, but Alexander noticed that his silence was linked to his poor posture. In the early twentieth century he developed a cure for his problem which focused on relearning correct posture.

Alexander teachers encourage their patients to move with the head leading and the spine following. These patterns of movement and posture are rehearsed repeatedly with a view to creating new motor pathways and improving posture, coordination and balance. Essentially, the mind is taught to modulate the autonomic nervous system through regular, supervised exercises.

The Alexander technique quickly became popular with performing artists. It was then noted that, apparently, it was also useful for a wider range of medical conditions. Today Alexander teachers claim that it is effective for treating asthma, chronic pain, anxiety and other illnesses.

Alexander teachers instruct their clients in a series of exercise sessions, each lasting up to an hour. They guide the process of relearning simple postures and body movements through a gentle, hands-on approach. As plenty of repetition is needed, at least 30 such sessions are usually required to master the technique. This obviously demands a considerable level of commitment from the client, in terms of both time and money.

Very little research has so far been conducted on the Alexander technique. Some promising findings have emerged in terms of improvement of respiratory function, reduction of anxiety, reduction of disability in Parkinson's disease, and improvement of chronic back pain. However, for none of these conditions (except perhaps chronic back pain) is the evidence sufficient to claim that the Alexander technique is of proven effectiveness. There are no serious risks associated with this method.

In short, the evidence for Alexander technique is not conclusive, but it might generate benefit for some health problems, provided patients are sufficiently committed and wealthy.

Alternative Diagnostic Techniques

Before administering a treatment, alternative therapists will often assess the patient's condition using a variety of diagnostic techniques. Some of these are entirely conventional, but others are not. The more unusual diagnostic techniques are often specific to a particular therapy, but the following list includes many diagnostic methods that are used in several disciplines:

- **Bioresonance:** electromagnetic radiation and electric currents from a patient's body are registered by an electronic device and used to diagnose everything from allergies to hormonal disorders. In treatment mode, the electrical signals are 'normalized' by the instrument and sent back into the patient's body.
- **Iridology**: each point on the iris is said to correspond to an organ, and irregularities are supposed to indicate problems with the corresponding organ.
- **Kinesiology**: muscle strength, tested manually, is claimed to be indicative of the health status of inner organs.
- **Kirlian photography**: high-frequency electrical currents applied to a patient's body generate electrical discharges which are turned into impressive, colourful images. These are in turn supposed to be indicative of human health.
- **Radionics**: a technique based on supposed energy vibrations in the body detected with pendula, divining rods or electrical devices.
- **Vega-test**: an electrodiagnostic device that can supposedly detect a range of conditions from allergies to cancer.

In nearly every case, these methods and the concepts behind them are not plausible, so their ability to diagnose accurately must be treated with great scepticism. Moreover, when these methods have been rigorously tested, the most reliable results of such investigations show that they are not valid. Finally, they typically fail the test of reproducibility, which means that ten practitioners generate ten different results.

These techniques are dangerous as they can generate false diagnoses. They can be misused by fraudulent practitioners to cause unwarranted fears in patients and to convince them to pay for ineffective or harmful treatments of conditions they did not have in the first place.

Alternative Diets

In alternative medicine, unsubstantiated health claims are being made for dozens of special diets that are not in line with accepted knowledge. Many of these are 'flavour of the month' approaches. To name but a few: Ama-reducing diet (Ayurvedic diet to burn off accumulated ama, which are supposed toxins); anthroposophic diet (lactovegetarian food with sour-milk products); Budwig's diet (fruit, juices, flaxseed oil and curd cheese); Gerson diet (fresh fruit juices, vegetables, supplements, liver extracts and coffee enemas to cure cancer); Kelly diet (anti-cancer diet including supplements and enzymes); Kousmine diet (anti-cancer diet with 'vital energy' foods, raw vegetables and wheat); macrobiotic diet (aimed at balancing yin and yang); McDougall diet (vegetarian diet, low fat, whole foods); Moerman diet (anti-cancer lactovegetarian diet with added iodine, sulphur, iron, citric acid and vitamins A, B, C, E); Pritikin diet (vegetarian diet combined with aerobic exercise); Swank diet (low amounts of saturated fat to combat multiple sclerosis).

Each of these diets has its own unique concept and is promoted for specific circumstances. Some must be followed long-term, others only until the condition in question is cured.

Clearly one would need to assess each diet on its own merits, yet little data has been gathered on any of those mentioned above or in general. Where evidence does exist, it is usually seriously flawed. For instance, the Gerson diet is promoted as a cancer cure, but the only positive evidence comes from an analysis which is now widely accepted to be fatally flawed and which should be ignored.

Several alternative diets can lead to malnutrition, particularly in seriously ill patients for whom it is important to consume a balanced diet with sufficient calorie intake. Feeding a highly restricted diet to a cancer patient, for instance, hastens death and reduces quality of life. Some proponents of these diets make patients feel guilty if they cannot follow their often tedious regimens. This can further reduce quality of life. Our advice is to stay well clear of alternative diets.

Alternative Exercise Therapies

The health benefits of regular exercise cannot be valued highly enough. Knowledge about exercise developed in all cultures, so unique exercise regimens emerged in different parts of the world and are often embedded in the specific concepts about health and disease of that region. Examples are t'ai chi (China) and yoga (India). Both include meditative aspects, need to be practised regularly and place a strong emphasis on disease prevention and wellbeing.

In addition to these traditional forms of exercise, there are modern variations on the theme. An example is pilates, developed relatively recently by Joseph Pilates (1880–1967). This approach integrates breathing, proper body mechanics and strengthening exercises, as well as stabilization of the pelvis and the trunk. It is estimated that there are now over 10 million people who practise pilates worldwide.

Although there is far less research into alternative exercises than into common sports or physiotherapy exercise, some encouraging conclusions have started to emerge. For example, yoga, which encompasses a whole lifestyle including diet and meditation, has been shown to be effective in reducing cardiovascular risks.

T'ai chi has also been studied quite thoroughly. It improves balance, prevents falls in the elderly, enhances cardiovascular fitness, increases joint flexibility, prevents osteoporosis in post-menopausal women and improves quality of life in patients suffering from chronic heart failure. There is, however, no significant evidence that alternative exercise therapies convey any additional benefits compared to many forms of conventional exercise.

In conclusion, regular exercise, whether exotic or conventional, is undoubtedly good for our health and wellbeing. Exercises are best learned in small groups and then have to be practised regularly – once or twice a week, or even daily. A well-trained, experienced tutor is important, as alternative exercise therapies can carry the sort of risks associated with any exercise that puts the body under strain.

Alternative Gadgets

An increasing number of alternative gadgets are being promoted with promises of health benefits for those who buy them. These gadgets have little in common except that the theories behind them conflict with mainstream science. For some entrepreneurs, alternative medicine is a highly profitable business, and there seems to be no limit to their inventiveness. They develop gadgets and claim that, if we buy and use them, our health will improve, certain illnesses will be cured or diseases will be prevented. The ideal medium for promoting these gadgets is, of course, the internet – there is no control over what claims can be made.

Examples of alternative gadgets are copper bracelets, devices that are said to shield us from electromagnetic radiation, jewellery with healing crystals, footbaths that supposedly extract toxins from our body, etc. In many cases, the only evidence provided by the manufacturer is statements by satisfied customers and 'experts', providing a thin veneer of credibility. Currently there are, for instance, dozens of websites where Professor Kim Jobst promotes the 'Q link' as a 'safe and effective tool that helps guard the cells of the body against electromagnetic field effects'. He also claims, 'Emerging evidence from early clinical cellular and molecular studies of the effects of Q link on cardiovascular, immune and central nervous systems are startling,' but this is simply not true.

The medicinal claims for these gadgets are often couched in apparently scientific language. This is to convince the consumer that the product is serious. On closer inspection, those with a scientific background can easily see their pseudo-scientific nature (i.e. the jargon is gobbledygook). The assumed mode of action of alternative gadgets is biologically implausible and no data exist to show that they have any positive health effects at all. Indeed, when devices have been tested, then the conclusion has invariably been disappointing.

The financial loss for patients is obvious, but there is also a health risk, as some people might employ these gadgets as alternatives to effective treatments. Seemingly harmless gadgets can then even hasten death.

Anthroposophic Medicine

Rudolf Steiner (1861–1925) developed a school of medicine based on imagination, inspiration and intuition. Anthroposophic medicine is influenced by mystical, alchemistic and homeopathic concepts and claims to relate to the spiritual nature of man. Rudolf Steiner created, among other things, the Waldorf schools, biodynamic farming and his own philosophy, known as anthroposophy.

Applying his philosophical concepts to health, he founded, together with Dr Ita Wegman, an entirely new school of medicine. It assumes metaphysical relations between planets, metals and human organs, which provide the basis for therapeutic strategies. Diseases are believed to be related to actions in previous lives; in order to redeem oneself, it may be best to live through them without conventional therapy. Instead a range of other therapeutic modalities is employed in anthroposophic medicine: herbal extracts, art therapy, massage, exercise therapy and other unconventional approaches.

The best-known anthroposophic remedy is a fermented mistletoe extract which is used to treat cancer. Steiner argued that mistletoe is a parasitic plant which eventually kills its host – a striking resemblance to a malignant tumour which also lives off its host and finally kills him/her. His conclusion, therefore, was that mistletoe can be used to treat cancer. The concepts of anthroposophic medicine are biologically implausible.

The efficacy of mistletoe preparations remains unproven – either for curing cancer or for improving the quality of life of cancer patients. Other elements of the anthroposophical concept are not well researched, and the therapeutic concept as a whole has so far not been rigorously tested.

Mistletoe injections have been associated with a range of adverse effects. The most important risk, however, is that of discarding conventional treatments. For example, anthropospohical doctors tend to advise parents against the immunization of their children, and some cancer patients forfeit conventional treatment for mistletoe extracts.

Anthroposophic medicine is biologically implausible, it has not been shown to be effective, and it is unlikely to work. It can also carry considerable risks.

Aromatherapy

Plant essences ('essential oils') have been used in several ancient cultures for treating or preventing illnesses or enhancing wellbeing, but the birth of aromatherapy proper was not until the publication of a book entitled *Aromathérapie* by the French chemist René Gattefosse in 1937. Gattefosse had previously burned his hand while working in his laboratory and immediately immersed it in lavender oil. To his amazement, the wound healed quickly without leaving a scar. This experience prompted him to study the medicinal powers of essential oils.

There are several ways of using essential oils. Most commonly, the diluted oil is applied to the skin via a gentle massage, but the oil can also be added to a bath or diffused in the ambient air. If combined with a massage, aromatherapy is definitely relaxing – but it is unclear whether the effect is caused by the oil, the gentle massage or both. Aromatherapists believe that different essential oils have different specific effects. Therefore therapists individualize these oils according to their patient's characteristics, symptoms, etc.

A consultation with an aromatherapist can last between one and two hours. The therapist will normally take a brief medical history, possibly conduct a short examination and then proceed by massaging a diluted essential oil into the skin of the patient. This usually is relaxing and agreeable. Aromatherapy is often advocated for chronic conditions such as anxiety, tension headache and musculoskeletal pain. Aromatherapists usually recommend regular sessions, even in the absence of symptoms, e.g. for preventing recurrences.

Some clinical trials confirm the relaxing effects of aromatherapy massage. However, these effects are usually shortlived and therefore of debatable therapeutic value. Some essential oils do seem to have specific effects. For instance, tea tree has antimicrobial properties. However, these are far less reliable than those of conventional antibiotics. The risks of aromatherapy are minimal, such as the possibility that some patients may be allergic to some essential oils.

In short, aromatherapy has short-term 'de-stressing' effects which can contribute to enhanced wellbeing after treatment. There is no evidence that aromatherapy can treat specific diseases.

Ayurvedic Tradition

'Ayurveda' means knowledge (veda) of life (āyus). It is one of the ancient Indian systems of healthcare and involves bringing about balance between body and mind. It includes individualized herbal remedies, diet, exercise (yoga), meditation, massage and other interventions. Health is perceived as a balance of physical, emotional and spiritual energies, and any deviation from health is thought to be caused by an imbalance of these elements. Treatment is aimed at re-establishing the balance through individualized prescriptions, usually of several interventions simultaneously.

Ayurvedic practitioners will take a medical history, examine the patient, diagnose the nature of the imbalance and try to restore balance. There is much emphasis on lifestyle advice, but Ayurvedic medicine supplements are also prescribed frequently. A consultation might take 30–60 minutes and numerous sessions are usually recommended, often lasting up to a year. All conditions are claimed to be treatable in the Ayurvedic tradition.

The whole system of Ayurveda has not been submitted to clinical trials, but elements of it have. The results are mixed. For instance, yoga has proven benefits for cardiovascular health. A recent trial of Indian massage, however, showed no positive effects in stroke patients. Ayurvedic remedies usually contain a multitude of herbal and other compounds. Some encouraging findings exist for conditions like acne, constipation, diabetes, chronic heart failure, obesity and rheumatoid arthritis. However, in none of these cases is the evidence sufficiently strong to warrant a positive recommendation.

Ayurvedic medicines have regularly been implicated for containing substances such as heavy metals. These can either be a contaminant or a deliberate addition – according to Ayurvedic belief, they generate positive medicinal effects if handled properly. In reality, however, heavy metals are highly toxic no matter how they are prepared.

In summary, ayurvedic healthcare is a complex system that cannot be easily evaluated. The current evidence suggests that some of its elements are effective while many others are essentially untested, or overtly dangerous, e.g. many herbal preparations.

Bach Flower Remedies

Edward Bach, who had worked as a microbiologist at the Royal London Homoeopathic Hospital in the early twentieth century, developed the concept of highly diluted plant infusions intended to cure emotional imbalances, which he thought to be the cause of all human illness. Inspired by the principles of homeopathy, Bach identified thirty-eight flower remedies, each one corresponding to an emotional disturbance such as depression, fear, loneliness or over-sensitivity. Administering the correct remedy, he believed, would cure the emotional disturbance and consequently the physical or psychological illness. For example, heather is used to treat self-centredness and honeysuckle is an antidote for those who live in the past. Similarly, wild rose is supposedly helpful in cases of apathy, and vervain treats the opposite problem of over-enthusiasm.

Flower remedies ('Bach Flower Remedies' is a brand name) are produced by placing the fresh flowers into spring water. Subsequently brandy is added to make up the actual remedies, which are similar to homeopathic remedies in that both are usually so dilute that no pharmacological action is conceivable. Also both schools claim to work through some sort of 'energy' transfer. However, homeopaths are adamant that flower remedies are fundamentally different: succussion (or shaking) is not part of the process of making flower remedies, and their prescription does not follow the 'like cures like' principle.

Flower remedies can be bought over the counter, but proponents argue it is best to consult a trained therapist. A therapist identifies the patient's underlying emotional imbalance, which, in turn, determines the choice of remedy. Flower remedies are also often recommended to healthy individuals with a view to preventing illness.

Several rigorous trials of flower remedies are available. None of them shows that this approach is effective beyond placebo in curing disease or alleviating symptoms. As the remedies are highly diluted, adverse effects are not likely.

Flower remedies are based on concepts which contradict current medical knowledge. The trial data fail to demonstrate effects beyond a placebo response. Therefore flower remedies are a waste of money.

Cellular Therapy

In conventional medicine, organs or cells are sometimes transplanted from one person to another, e.g. bone-marrow transplantations or blood tranfusions. This is entirely different from cellular therapy as used in alternative medicine, sometimes also called 'live cell therapy' or 'cytotherapy'.

In 1931 the Swiss surgeon Paul Niehans had the idea of injecting preparations from animal foetuses into humans for the purpose of rejuvenation. This concept seemed plausible to lay people and many influential individuals who could afford this expensive treatment became Niehan's patients. When it emerged that Niehan's *Frischzellen Therapie* (fresh cell therapy) was dangerous – thirty deaths had been reported by 1955 – his preparations were banned in several countries.

Meanwhile several similar cellular treatments had emerged, particularly on the European continent. Examples include 'Thymus' therapy (injection of the extracts from the thymus gland of calves) or 'Ney Tumoin' (protein extracts from calves or cows) or 'Polyerga' (protein extracted from pig spleen) or 'Factor AF2' (extract from spleens and livers of newborn sheep). These preparations are usually injected by doctors (non-doctor therapists are not allowed to give injections in most countries) who claim that they have anti-cancer properties, stimulate the immune system or simply regenerate organs or rejuvenate the body in a general sense.

Thymus therapy has been extensively researched as a cancer treatment. The totality of this evidence does not show the approach to be effective. Other preparations have either generated similarly negative results or have not been submitted to clinical trials. However, it is known that any treatment that introduces foreign proteins directly into the bloodstream can lead to anaphylactic shock, the most serious type of allergic reaction. If this condition is not treated adequately and immediately, it can result in death.

The seemingly plausible principle of cellular therapy continues to appeal to the rich and super-rich. None of the claims of cellular therapy are, however, supported by scientific evidence, so these treatments are both dangerous and a waste of money.

Chelation Therapy

Chelation therapy started as a branch of conventional medicine to remove heavy metals and other toxins from the body by introducing powerful chemical agents, which bind to the toxins and are subsequently excreted. This conventional form of chelation therapy is indisputably effective and often life-saving. In alternative medicine, chelation therapy is used in very different ways and has two main applications.

First, alternative therapists use chelation to remove toxins, but the source of these toxins is unclear. For example, they may attempt to remove mercury which allegedly leaked from dental fillings or vaccines. There is, however, no evidence to suggest any toxicity from these sources. Thus chelation therapy is employed to fix a non-existing problem.

Second, chelation therapy is used for eliminating calcium ions from the blood, based on the notion that calcium deposits in the arterial wall are responsible for arteriosclerosis which, in turn, is seen as the cause of heart disease, stroke, peripheral arterial disease and other conditions. Consequently, alternative chelation therapists insist that their treatment is helpful for coronary heart disease, stroke prevention, peripheral vascular disease and a range of conditions from arthritis to osteoporosis.

Alternative chelation therapists usually advocate a whole series of treatments. In total, this can cost the patient thousands of pounds.

The claims that chelation is effective for coronary heart disease, stroke or peripheral arterial disease is based on outdated scientific theories. Chelation therapy has been tested repeatedly, but these clinical trials fail to demonstrate effectiveness. Serious adverse effects, including deaths due to electrolyte depletion, have been associated with chelation therapy. In fact, there have been nine documented deaths since 1990.

Chelation therapy, as used in alternative medicine, is disproven, expensive and dangerous. We urge patients not to use this treatment.

Colonic Irrigation

The notion that we are poisoning ourselves with toxic intestinal waste products from ingested food seems plausible to many lay people and is therefore widespread. It forms the basis for a range of alternative approaches which allegedly free the body of such 'auto-intoxication'.

One of them is colonic irrigation, or colon therapy as it is also called, whereby enemas are used to 'cleanse the body'; sometimes herbs, enzymes or coffee are added to the water which is administered via the rectum. The popularity of this treatment can be explained through its apparently logical concept and through its continuous promotion by the popular media and certain celebrities.

A treatment session involves partial undressing, insertion of a tube via the rectum and receiving considerable amounts of fluids via this route. The fluid is later extracted and, on closer inspection, appears to be loaded with 'waste products'.

This visual impression helps to convince patients that colonic irrigation achieves what it claims: the elimination of residues that the body is best rid of. Treatment might last for approximately 30 minutes and long-term therapy is sometimes advised, with weekly or twice weekly sessions. Colonic irrigation is promoted as a treatment for gastrointestinal disorders, allergies, obesity, migraine and many other chronic illnesses.

Enemas have an undoubted role in conventional medicine. The use of colonic irrigation as employed in alternative medicine is, however, an entirely different matter. None of the waste products of our body 'poison' us; they are eliminated through a range of physiological processes, unless we are suffering from severe organ failure.

There is no reliable clinical evidence that colonic irrigation does any good at all and some evidence that it causes serious harm by, for example, perforating the colon or depleting our body of electrolytes.

Colonic irrigation is unpleasant, ineffective and dangerous. In other words, it's a waste of money and a hazard to our health.

Craniosacral Therapy (or Cranial Osteopathy)

William G. Sutherland, who practised as an osteopath in the 1930s, became convinced that our health is governed by minute motions of the bones of the skull and sacrum. These subtle rhythms, he believed, are fundamental to the self-healing processes of our bodies. Craniosacral therapy aims at restoring the rhythmic motions when they are restricted. This is claimed to help with a range of conditions, particularly in children: birth trauma, cerebral palsy, chronic pain, dyslexia, headaches, learning difficulties, sinusitis, trigeminal neuralgia and many others. Today, craniosacral therapy is practised by several alternative therapists, including osteopaths, chiropractors, naturopaths and massage therapists.

A consultation with a craniosacral therapist would include a detailed manual diagnosis, so the first session may last one hour or longer. Subsequent therapeutic sessions, during which the therapist gently manipulates the skull and sacrum, would be shorter. A typical treatment series might involve six or more sessions.

Conventional wisdom has it that, during early childhood, the bones of the skull and the sacrum fuse to form solid structures. Even if minute motions between these bones were possible, they would be unlikely to have a significant impact on human health. In other words, craniosacral therapy is biologically implausible.

The little research that exists fails to demonstrate that craniosacral therapy is effective in treating any condition. Moreover, therapists struggle to give consistent diagnoses for the same patient, probably because they are attempting to detect a non-existent phenomenon. Mothers bringing their children to a therapist are sometimes impressed by the positive reaction. This is likely to be a relaxation response caused by the gentle touch and calming manner of the therapist, but these effects are usually shortlived and the treatment can be expensive. There are no conceivable risks, but if severely ill children are treated with craniosacral therapy instead of an effective treatment, the approach becomes life threatening.

Crystal Therapy

Crystals such as quartz or other gemstones are sometimes used to enable 'energy healing'. Crystal therapists claim that crystals can move, absorb, focus and diffuse healing 'energy' or 'vibrations' within the body of a patient. This, in turn, is said to enhance the self-healing ability of the patient. Illness allegedly occurs when the individual is misaligned with the 'divine energy' that is 'the foundation of all creation'. The approach is not in accordance with our understanding of physics, physiology or any other field of science. Crystal therapy therefore lacks scientific plausibilty.

A treatment session typically involves the fully clothed patient lying down. The therapist then intuitively identifies problem areas such as blockages of energy flow and places crystals over them to restore flow or re-establish balance. One treatment typically lasts for 30–60 minutes.

Crystal therapy is normally used by patients as an addition to conventional treatments. It is employed in the treatment of virtually all medical conditions, for improving the quality of health in individuals or for disease prevention. Therapists sometimes use crystal wands as part of aura therapy in order to cleanse a patient's aura.

Patients who believe in crystal therapy will often buy their own crystals and carry them about their person in order to treat minor conditions. Although healing crystals can be relatively cheap, healing crystal jewellery sometimes costs several hundred pounds.

There is no evidence that crystal therapy is effective for any condition. The positive effects experienced by some patients are almost certainly due to expectation, relaxation or both.

Similarly, there is no evidence that the carrying or wearing of crystals is effective for any condition. If used as an alternative to life-saving treatments, crystal therapy would be life threatening, but there are no conceivable direct risks in this approach.

Cupping

Cupping involves stimulating certain points of the body surface by attaching cups that generate suction. Cupping is an ancient treatment that has been practised in places such as China, Vietnam, the Balkans, Russia, Mexico and Iran. Essentially, the air in a glass cup is heated over a flame and the cup is then swiftly placed on the skin. As the air in the cup cools down, a vacuum develops which creates suction. This is visible as the skin and its underlying soft tissue are partly sucked into the cup. Sometimes the skin is lacerated beforehand, and the suction then draws blood from the cutaneous microcirculation. This form of cupping was popular in connection with bloodletting in Europe.

In traditional Chinese medicine, cupping can be used as one of several ways of stimulating acupuncture points. Hence, Chinese cupping has the same underlying philosophy as acupuncture.

Cupping is used to treat a variety of conditions, such as musculoskeletal problems, asthma or eczema. Some practitioners even claim to treat conditions such as infertility, influenza and anaemia. Usually it is employed in combination with other therapies. The actual treatment lasts about 20 minutes and repeat sessions are usually advised. Cupping is practised by a range of alternative practitioners including naturopaths, acupuncturists and chiropractors.

The only controlled trial of cupping did not demonstrate the effectiveness of this therapy in reducing pain. However, the cupping procedure and its visible aspects (e.g. skin being sucked into the cup as if by 'magic') are likely to generate an above-average placebo response.

When handled correctly, there are few risks. On the other hand, the sucking action can leave round bruises which can last for several days. There was a very public demonstration of this in 2005, when the actress Gwyneth Paltrow attended a New York film premiere wearing a backless dress and showing dark bruises across her shoulders. Also, the bloodletting version of cupping carries the risk of infection.

Cupping has a long history but there is no evidence that it generates positive effects in any medical condition.

Detox

Detox or detoxification is the elimination of accumulated harmful substances from the body. Conventional detoxification has its established place in medicine, e.g. for eliminating poisons that have been ingested or injected. The term is also used for weaning addicts off drugs or alcohol. In alternative medicine, however, detox has been hijacked and has acquired a slightly different meaning. It is suggested that either the waste products of our normal metabolism accumulate in our body and make us ill, or that too much indulgence in unhealthy food and drink generates toxins which can only be eliminated by a wide range of alternative treatments.

Detox is often recommended after periods of over-indulgence, e.g. after the Christmas holiday. It is incessantly promoted by magazines and certain celebrities. In alternative medicine, detox can mean anything from a course of self-administered treatments to a week in the luxury of a health spa. The former, for example, might consist of a mixture of herbal and other supplements or several days of dieting, which costs just a few pounds. The latter, however, can cost a few hundred pounds.

The conventional form of detoxification can be life-saving. In alternative medicine, however, detox is a scam. Supporters of alternative detox have never demonstrated that their therapies are able to reduce levels of toxins. This would be very easy to achieve, e.g. by taking blood samples and measuring blood levels of certain toxins. In any case, the human body is well equipped with highly efficient organs (liver, kidney, skin) to eliminate 'toxins' due to over-indulgence. Drinking plenty of water, gentle exercise, resting and eating sensibly would rapidly normalize the body after a period of over-indulgence. An expensive detox is not required to achieve this aim.

Detox, as used in alternative medicine, is based on ill-conceived ideas about human physiology, metabolism, toxicology, etc. There is no evidence that it does any good and some treatments, such as chelation or colonic irrigation (see separate entries in this Rapid Guide), can be harmful. The only substance that is being removed from a patient is usually money.

Ear Candles

Allegedly, ear candles were used in China, Egypt, Tibet, by the Hopi Indians in America, and even in Atlantis!

Ear-candling entails placing a hollow candle into the ear of the patient and lighting the far end of the candle, which then burns slowly over about 15 minutes. This generates mild suction and is supposed to stimulate energy points. Thereafter, the candle is extinguished and the content of the near end of the candle is usually displayed for inspection. Many therapists inform their patients that the remnant left behind at the end of treatment is ear wax, suggesting that it has been drawn out of the ear through the 'chimney effect' produced by the burning candle.

Ear-candling is used for the removal of ear wax and for the treatment of hay fever, headaches, sinusitis, rhinitis, colds, influenza and tinnitus. It is even claimed candling can lead to 'sharpening of mental functioning, vision, hearing, smell, taste and colour sensation'.

There is no shortage of anecdotes published to promote the use of ear candles. However, a series of experiments concluded that ear candles do not eliminate any substance from the ear.

A study conducted in 1996 by Spokane Ear, Nose, and Throat Clinic in America showed that a burning candle does not produce any negative pressure at all, and that the deposit is, in fact, candle wax. Indeed, the same group of researchers also demonstrated that instead of removing ear wax, ear candles leave a deposit of wax in volunteers who, prior to the experiment, had no ear wax.

There is no evidence that candling is effective in the treatment of any of the other conditions mentioned above.

Ear candles are not free of risks: burns, occlusion of the ear canal and perforations of the eardrum are on record. There are also cases of house fires resulting from candling sessions.

In light of these risks and the lack of evidence, ear-candling is a treatment that should be avoided.

Feldenkrais Method

Moshe Feldenkrais (1904–1984) was a physicist and electro-engineer who suffered badly from chronic knee pain. No treatment he tried helped and he thus decided to develop his own cure.

The Feldenkrais method is based on the belief that body and spirit form a fundamental whole. The founder declared, 'I believe that the unity of mind and body is an objective reality. They are not just parts somehow related to each other, but an inseparable whole while functioning.' Feldenkrais published his first book outlining his philosophy in 1949 – *Body and Mature Behavior: A Study of Anxiety, Sex, Gravitation and Learning.*

The therapy is carried out in two steps: during the 'functional integration' phase, the practitioner uses touch to demonstrate to the patient techniques that improve breathing and body movement. During the subsequent 'awareness through movement' phase, the practitioner teaches the patient to correct so-called false movements.

The aim is to improve everyday functions. According to Feldenkrais, behaviour is not innate, but is merely acquired. False behaviour, he thought, was 'a groove into which a person sinks never to leave unless some physical force makes him do so'. His treatment, he was convinced, provides that force.

The therapy consists of a series of sessions usually carried out in small groups. Once the lessons are learned, the patient has to practise the techniques continuously at home. The conditions treated include musculoskeletal problems, multiple sclerosis and psychosomatic problems.

Only about half a dozen rigorous clinical trials are currently available. Their results are far from uniform. Some, but not all, results suggest that the Feldenkrais method is useful for multiple sclerosis patients. For other conditions, the evidence is even less convincing. There are no conceivable serious risks.

The Feldenkrais method is not well researched, and the existing trials do not provide compelling evidence that it is effective for any condition.

Feng Shui

Chinese medicine assumes that all health is governed by the flow of energy (Ch'i) and the balance of yin and yang within the body, but these concepts can also be applied to the things that surround us. Feng shui is the Chinese art of placing objects in accordance with the theory of yin and yang in order to optimize the flow of life energy, which, in turn, is thought to influence health and wellbeing. Feng shui consultants give advice on the position of objects in an office or home. They may, for instance, place a screen in a certain position to make sure that the energy is travelling in the right direction, or they might advise their clients to reposition their beds so that they can benefit from the right energy flow while sleeping.

Feng shui is not biologically plausible, because its basic tenets make no sense in the context of modern science. The benefit some people experience after following the advice of feng shui consultants could be due to expectation and has no physiological basis, nor is it likely to last.

Feng shui consultants do not normally claim to cure diseases, but they say that their work can improve wellbeing and prevent ill-health. Increasingly, feng shui consultants are giving advice on how to deal with the health effects of electromagnetic fields in the home, even though there is no evidence that such fields are harmful. These consultants generally charge considerable amounts of money for their services.

It would not be difficult to test some of these claims, but as yet there have been no serious studies. However, informal tests comparing the judgements of feng shui consultants demonstrate significant conflicts over their interpretation of the energy flow in any given space, which implies that their advice is based on subjective imaginings.

Therefore all we can say is that there is no evidence to show that feng shui does anything but enrich those who promote it. A competent interior designer can probably offer equally good, if not better, advice.

Food Supplements

Food supplements are substances usually taken by mouth to increase the intake of vitamins, minerals, fats, amino-acids or other natural substances in order to maintain or improve health, fitness or wellbeing. At present, sales are booming. Regulation of food supplements varies nationally but, generally speaking, it is very lax. Manufacturers can sell supplements without providing proof that they do any good at all, and often without sufficient data on safety.

Medical claims are not normally allowed for food supplements. This does not, however, stop the industry from cleverly conveying the message that this or that supplement is effective for treating this or that condition. Health writers, books and the internet relentlessly target the consumer with exactly that aim.

It is obvious that not all food supplements are the same, as is the case with herbal supplements, which were discussed in detail in Chapter 5. Some are likely to be useful and are supported by evidence; others are either unproven or disproven, and many supplements carry risks of adverse effects. The dangers can be due to a supplement's inherent properties, contamination (e.g. heavy metals), or adulteration (e.g. synthetic drugs). Also, it is likely that there are many unknown adverse effects due to lack of research and under-reporting.

Fish-oil capsules, as discussed in Chapter 6, are an excellent example of an effective supplement, because they have been proved to reduce the risk of heart disease. They may also reduce inflammation, which might make them beneficial for rheumatoid arthritis and many other conditions.

Shark cartilage, also discussed in Chapter 6, is an example of a popular supplement which has been shown to be ineffective. Although it is probably harmless, it can distract patients from seeking more appropriate treatment and it is certainly damaging for the sharks who are victims of the supplement industry.

Vitamin B_6 is an example of a supplement that can be harmful in large doses. It can result in nerve damage to the arms and legs. There are several reports of people reporting such complications having taken 500mg of B_6 per day.

Hypnotherapy

Hypnotherapy is the use of hypnosis, a trance-like state, for therapeutic purposes. It has a long history and can be traced back to ancient Egypt, but its modern development started in the eighteenth century with the work of the charismatic Viennese scientist Franz Mesmer. He was followed in the nineteenth century by the Scottish physician James Braid.

In recent years, hypnotherapy has become recognized in several areas of healthcare. Hypnotherapists treat a range of chronic conditions, including pain, anxiety, addictions and phobias. Hypnotherapy is practised by several healthcare professionals, including psychologists, counsellors and doctors. One session lasts 30–90 minutes and, depending on the condition and the responsiveness of the patient, 6–12 sessions are normally recommended. Autogenic training is a self-hypnotic technique, which, after some instruction, can be practised without the help of a therapist.

People who are suggestible generally respond best. Dozens of clinical trials show that hypnotherapy is effective in reducing pain, anxiety and the symptoms of irritable bowel syndrome. However, according to reliable reviews by the Cochrane Collaboration, it is not effective for smoking cessation, even though it is frequently promoted in this context. There is much less research for autogenic training, but the existing evidence is encouraging for anxiety, stress, hypertension, insomnia and some pain syndromes.

Hypnotherapy and autogenic training are relatively safe, but they should not be used by people with psychoses or other severe mental problems. With hypnotherapy, the recovery of repressed or false memories can create problems, and cases of false-memory syndrome (i.e. remembering distressing events which, in reality, have never occurred) have been reported.

The prudent use of hypnotherapy can be helpful for some patients. Whether this is a specific effect of the treatment or a non-specific (placebo) effect is difficult to say. Autogenic training has the added advantage of being an economical self-help approach that maximizes each patient's own involvement. Neither treatment is associated with serious risks when applied correctly.

Leech Therapy

Hirudo medicinalis is a small, black, worm-like animal capable of attaching itself to humans or animals, sucking substantial amounts of blood from the skin. During this process, it increases considerably in size and finally, when it is saturated with blood, it drops off.

Leeches were used for medicinal purposes in ancient Babylon, and in more recent centuries in Europe as a mode of bloodletting, as discussed in Chapter 1. Today their only use in conventional medicine is in plastic surgery: clinical trials show that applying leeches post-operatively improves the cosmetic results of some operations.

In alternative medicine, leeches are employed for a range of conditions. Some therapists believe that they eliminate toxins from the body; others use them to treat painful local conditions such as osteoarthritis.

While sucking blood, leeches inject pharmacologically active substances into the body. Initially they inject an anaesthetic substance which enables them to bite through the skin without causing pain. Subsequently they excrete a substance that prevents blood from clotting so that they can suck blood easily. This substance is called hirudoine and is a well-researched anticoagulant. It can now be synthesized and is used widely in mainstream medicine for its anticoagulation properties.

A German group recently published a series of clinical trials which suggested that the application of several leeches over the knee eases the pain of osteoarthritis. These studies still await independent replication. All other claims of the alternative use of leech therapy are unsupported by evidence.

If done properly, there are few risks. Many patients, however, might feel uncomfortable about leech therapy, either because of the aesthetic aspects of the treatment or the fact that, after one session, the animals are usually destroyed.

In summary, leeches have a long history of medical use. There is some evidence that their use is effective in reducing the pain of osteoarthritis of the knee. There is no evidence to support their use for any other treatment performed by an alternative therapist.

Magnet Therapy

Nowadays, rapidly fluctuating magnetic fields are employed in conventional medicine in hi-tech imaging instruments and for promoting the healing of bone fractures. However, alternative medicine tends to focus on the use of static magnets, which give rise to a permanent magnetic field.

Such static magnets have always attracted the interest of physicians, but the boom in magnet therapy began in Europe in the eighteenth century. Although static magnets fell out of favour as medicine progressed, a plethora of static magnets are today again popular within alternative medicine. Their use is promoted for many conditions, most frequently to alleviate chronic pain. These static magnets are worn as wrist bands, belts, leg wraps, shoe inlays, patches, etc. Magnetic mattresses or seat covers are also available. The magnetic strength of these devices varies between 10 and 1,000 Gauss. Static magnets can be purchased through numerous outlets and, more often than not, the consumer/patient would not have had any contact with a healthcare practitioner.

Subtle effects of magnetic fields are observable, for instance, on cell cultures. The question is whether these translate into any therapeutic benefit.

Most of the clinical research on static – meaning constant field strength rather than fluctuating strength – magnets relates to pain control. Researchers at Exeter University recently included nine placebo-controlled, randomized trials in a meta-analysis. The results do not support the use of static magnets for pain relief. For other problems, such as menstrual symptoms or varicose veins, the evidence is equally unconvincing.

Static magnets are unlikely to cause direct adverse effects. As they are usually self-administered, there is a danger of missing serious diagnoses and losing valuable time for early treatment of serious diseases.

Static magnets are popular, and the market is booming, but it is important to realize that there is no evidence that they offer any medical benefit, and indeed there is no reason why they ought to. There is more information about magnet therapy in Chapter 5.

Massage Therapy

Massage is as old as medicine itself. Today many variations exist; for example, classical 'Swedish' massage focuses on muscular structures. Other forms of massage treatment include:

- **Bowen therapy**: gentle soft-tissue technique influencing the nervous system.
- **Lymphatic drainage**: massaging along lymph channels to enhance lymph flow.
- **Marma massage**: traditional Indian massage.
- **Myofascial release**: technique reducing tension in fascia and connective tissue.
- **Relaxation massage**: gentle superficial techniques.
- **Rolfing**: forceful massaging where the therapist's whole body applies pressure.
- **Sports massage**: muscular techniques adapted for the needs of athletes.

While many massage therapies are based on a sound understanding of anatomy, some rely on unproven and unlikely philosophies. These more exotic forms of massage therapies include shiatsu, craniosacral therapy and reflexology (which are all discussed elsewhere in this appendix), as well as polarity massage (balancing positive and negative energy), trigger-point massage (pressurizing trigger points aimed at reducing local pain or influencing the function of distant organs) and acupressure (pressurizing, rather than needling, acupuncture points).

Massage is practised by specialist massage therapists, physiotherapists, nurses, alternative practitioners of all types and other healthcare professionals. They aim to treat both physical problems (e.g. musculoskeletal pain) and psychological conditions (e.g. anxiety or depression).

There is encouraging evidence that massage is beneficial for some musculoskeletal problems, especially back pain, anxiety, depression and constipation. It acts, possibly, by increasing local blood flow and releasing endorphins in the brain. Adverse effects are rare.

Generalizing is problematic, but massage is probably effective for some conditions and improves wellbeing in most patients. The more exotic forms of massage are generally unlikely to offer any extra benefit.

Meditation

Meditation covers a range of techniques that direct the subject's attention to a symbol or sound or thought in order to achieve a higher state of consciousness. Most religions have developed techniques that bring about altered states of consciousness. They may include repeating a mantra or listening to one's own rhythm of breathing. Such rituals can lead to a deep relaxation and mental detachment. This 'relaxation response' can also be used therapeutically for reducing stress, which, in turn, can bring about other health benefits such as lowering blood pressure or pain control. Meditation is normally taught in a series of sessions; subsequently patients who have mastered the technique are requested to practise on a daily basis.

During the meditative state, a range of physiological functions are altered. For example, respiratory rate and heart rate are slowed, and brain activity is reduced. Proponents of meditation claim to treat conditions such as anxiety, hypertension, asthma or drug dependency.

Certain forms of meditation (e.g. Transcendental Meditation) have strong religious associations and may be part of larger systems of beliefs and practices that patients may not feel is appropriate. For example, Hinduism is the most ancient religion to advocate meditation as a spiritual practice. 'Mindfulness Meditation' is an approach which has been developed purely for therapeutic purposes and does not raise such issues.

Research into meditation is scarce and often seriously flawed. Truly independent evaluations are rare. Nevertheless, it seems likely that meditation offers many of the benefits associated with relaxation. Some alternative therapists suggest that meditation can have a direct impact on serious conditions, such as cancer, but there is no evidence to support such claims.

Some reports suggest that mental illnesses can be exacerbated through meditation, so patients with such problems should not use it.

In conclusion, meditation can be relaxing and thus increase well-being. In this way it can prove to be useful for many people. In the absence of mental illness, it seems to be a safe form of therapy.

Naturopathy

The naturopathy movement began in eighteenth-century Europe, where people such as the priest Sebastian Kneipp preached the value of curing disease with the means that nature has provided. Naturopaths are convinced of nature's own healing power (vis *medicatrix naturae*), a gift that all living organisms are believed to possess. In their view, ill-health is the result of disregarding the simple rules of a healthy lifestyle. Therefore much emphasis is put on a good diet, regular exercise, sufficient sleep, etc. Once an illness occurs, naturopaths employ herbs, water cures, massage, heat, cold and other natural means to cure it.

Consulting a naturopath is similar to seeing a conventional doctor, inasmuch as a diagnosis will be made by taking the patient's history and a physical examination. The main difference lies in the nature of the prescriptions. Naturopaths do not prescribe synthetic drugs. Their treatment usually consists of the treatments mentioned above plus lifestyle advice. Naturopaths tend to treat chronic benign conditions such as arthritis and headache.

Even though it would be feasible to test the effectiveness of the whole naturopathic approach, such trials are so far not available. However, much of the naturopathic approach is eminently valid (e.g., healthy diet, regular exercise). Similarly, certain herbal remedies are of proven value (see Chapter 5).

On the other hand, naturopathy can carry risks, particularly if it delays a patient with a serious condition from seeking urgent conventional treatment. Indeed, many naturopaths are against mainstream medicine and advise their patients accordingly – for instance, many are not in favour of vaccination. Also, some naturopathic treatments, such as particular herbal remedies, can carry risks.

A general judgement about the wide variety of naturopathic treatments is not possible. Each naturopathic treatment must be critically assessed on its own merits, and it is likely to be covered elsewhere in this appendix. For any serious condition, naturopathy should not be seen as an alternative to conventional medicine.

Neural Therapy

Neural therapy uses injections with local anaesthetics for the identification of health problems, treatment of diseases and alleviation of symptoms. The brothers Ferdinand and Walter Huneke were German doctors who pioneered neural therapy during the first half of the twentieth century. They made observations regarding the local anaesthetic Novocain which convinced them that injecting this drug around a 'field of disturbance' (*Störfeld*) generates dramatic effects in other parts of the body. This, they postulated, has nothing to do with the pharmacological action of the local anaesthetic, but is mediated through the autonomic nervous system.

One key event, for instance, was when Huneke injected Novocain into the skin around a leg wound of a patient who then was cured of an old shoulder pain within seconds. This type of observation was called *Sekundenphänomen* (phenomenon of a cure within seconds).

The brothers Huneke claimed the 'fields of disturbance', often old scars, injuries or sites of inflammation, can exert strong influences throughout the body which in turn can cause problems in distant body structures. Treating a particular problem may involve injecting Novocain or other local anaesthetics into one or two sites that may be 'fields of disturbance'. When the correct site is located, the problem is cured.

Neural therapy is particularly popular in German-speaking countries. There are also many practitioners in the Spanish-speaking world, largely thanks to its promotion in the 1950s by a German-Spanish dentist called Ernesto Adler.

Injecting local anaesthetics into an area of pain will reduce that pain – but that is a predictable pharmacological effect and not what neural therapy is about. The concepts of neural therapy have little grounding in science, and the few clinical trials that exist have not produced any convincing evidence to support neural therapy. Occasionally, the local anaesthetic drug can cause adverse reactions, but such incidents are rare.

Although the injection of local anaesthetics as performed by many doctors can control pain, neural therapy is biologically implausible and is not backed up by sound evidence.

Orthomolecular Medicine

'Orth' means correct, and orthomolecular medicine (also known as optimum nutrition) means administering doses of vitamins, minerals and other natural substances at levels that have to be exactly right for the individual patient. Proponents of this approach believe that low levels of these substances cause chronic problems which go beyond straightforward mineral or vitamin deficiency. These problems include a tendency to suffer from infections such as the common cold, lack of energy or even cancer. This means each patient is initially assessed to determine precisely which substances he or she needs. Subsequently the 'correct' mixture is prescribed. The hallmarks of orthomolecular medicine are the extremely high doses that are usually suggested and the individualization of the prescription.

Some of the diagnostic methods that are being used for defining the right mixture of substances are not of proven validity. For instance, hair analysis is often employed, yet it generates spurious results in this context. The medicinal claims made are neither plausible nor supported by data from clinical trials. Thus there is no evidence that orthomolecular medicine is effective.

Proponents would strongly dispute this statement and refer to a plethora of studies that show the efficacy of vitamins. After all, vitamins are substances that are vital for humans – without them we cannot survive. However, our normal diet usually provides sufficient vitamins and the treatment of vitamin deficiencies is unrelated to the specific principles of orthomolecular medicine.

In excessive doses, vitamins can cause harm. Virtually all of these substances will cause adverse effects if grossly overdosed over prolonged periods of time – and this is precisely what is recommended by proponents of orthomolecular medicine.

In summary, the concepts of orthomolecular medicine are not biologically plausible and not supported by the results of rigorous clinical trials. These problems are compounded by the fact that orthomolecular medicine can cause harm and is often very expensive.

Osteopathy

Osteopaths offer a manual therapy involving a range of techniques, particularly mobilization of soft tissues, bones and joints. The American Andrew Taylor Still founded osteopathy in 1874 – around the time when chiropractic therapy (see Chapter 4) was created by D. D. Palmer. Osteopathy and chiropractic therapy have much in common, but there are also important differences. Osteopaths tend to use gentler techniques and often employ massage-like treatments. They also place less emphasis on the spine than chiropractors, and they rarely move the vertebral joints beyond their physical range of motion, as chiropractors tend to do. Therefore osteopathic interventions are burdened with less risk of injury.

In the US, doctors of osteopathy (DOs) are entirely mainstream and only rarely practise manual therapies. In the UK, osteopaths are regulated by statute but considered to be complementary/alternative practitioners. British osteopaths treat mostly musculoskeletal problems, but many also claim to treat other conditions such as asthma, ear infection and colic.

There is reasonably good evidence that the osteopathic approach of mobilization is as effective (or ineffective) as conventional treatments for back pain. For all other indications, the data are not conclusive. In particular, the overall conclusion from several clinical trials is that there is no good evidence to support the use of osteopathy in non-musculoskeletal conditions.

Because their techniques are generally much gentler than those of chiropractors, osteopaths cause adverse effects much less frequently. Nevertheless, people with severe osteoporosis, bone cancer, infections of the bone or bleeding problems should confirm with the osteopath that they will not receive forceful manual treatments.

The evidence that the osteopathic approach is effective for treating back pain is reasonably sound. If, however, you receive no significant benefit then be prepared to switch to physiotherapeutic exercise, which is backed by similar evidence and which can be done in groups and therefore is more cost-effective.

Oxygen Therapy

Oxygen is essential for life and has many uses in conventional medicine. For instance, if the lungs are no longer capable of taking up sufficient amounts of oxygen, the patient may be given oxygen-enriched air to breathe.

In the context of alternative medicine, however, oxygen therapy is much more controversial. Alternative oxygen therapy is practised in a variety of ways, which differ according to the way in which oxygen is administered, the type of oxygen (e.g. oxygen [O_2] or ozone [O_3]) administered or the conditions being treated.

There are many ways to administer oxygen. For example, it can be injected subcutaneously or a patient's blood can be drawn, exposed to oxygen and re-injected into the body. Alternatively, oxygen-enriched air can be applied to the skin, or oxygen-enriched water can be used for colonic irrigation.

The range of conditions supposedly treated by oxygen therapy includes cancer, AIDS, infections, skin diseases, cardiovascular conditions, rheumatic problems and many other illnesses.

The fact that we all need oxygen for survival does not mean that more oxygen than normal is beneficial. Indeed, this is certainly not the case: there is plenty of evidence that an excess of oxygen can be harmful to patients. And, of course, ozone is well known for its extreme toxicity.

Some of the many forms of oxygen therapy have been tested in clinical trials. The results were not convincing and it is therefore safe to say that no type of alternative oxygen therapy is supported by sound evidence. Thus the potential risks clearly outweigh the documented benefits.

The conclusion is clear. Oxygen has a wide variety of uses in conventional medicine, but its role in alternative medicine is based on biologically implausible theories. Therefore, alternative oxygen therapy is unproven and, worse still, potentially harmful. We recommend avoiding it.

Reflexology

Manual massages of the feet are usually experienced as relaxing and it is therefore not surprising that they were used in various ancient cultures. But reflexology is different. It is based on assumptions by William Fitzgerald who, in the early twentieth century, postulated that the body is divided into ten vertical zones, each represented by part of the foot. Fitzgerald developed maps of the soles of the feet showing which areas correspond to which inner organs.

Reflexologists take a brief medical history and then manually investigate the foot. If they feel a resistance in one area they are likely to diagnose a problem with the corresponding organ. The therapy then consists of a high-pressure massage at this point, which is believed to repair the function of the troubled organ and ultimately to improve the patient's health or prevent illness. One session may last half an hour, and a series of treatments may consist of ten or more sessions. In the absence of any health problems, many therapists recommend regular maintenance sessions for disease prevention.

The postulated reflex pathways between a certain area of the foot and an inner organ do not exist, and the notion that resistance in one area of the foot is a reliable indicator for a problem with a certain organ (e.g. kidney) is unfounded. Hence, the technique is not biologically plausible. Moreover, several different versions of reflexology maps exist – reflexologists cannot even agree among themselves how to apply the treatment. Clinical trials have shown that reflexology has no diagnostic value. Its effectiveness in treating certain health problems has been tested repeatedly. Even though the results have not been uniform, they generally do not demonstrate convincingly that this therapy is effective. There is also no evidence that regular reflexology might prevent diseases.

People with bone or joint conditions of the feet or lower legs might be harmed by the often forceful pressure applied during treatment. Otherwise no serious risks are known.

The notion that reflexology can diagnose health problems has been disproved and there is no convincing evidence that it can treat any condition. Reflexology is expensive and offers nothing more than could be achieved from a simple, relaxing foot massage.

Reiki

Reiki is a system of spiritual healing or 'energy' medicine which is similar to the laying on of hands. Reiki healers believe in the existence of a universal energy which they can access in order to generate healing effects in humans, animals and plants. This universal energy flows through a reiki healer's hands when he or she places the palms upon or close to the recipient. This allegedly enhances the recipient's own healing potential. However, the concepts of reiki are contrary to our understanding of the laws of nature. The approach therefore lacks biological plausibility.

Reiki is popular far beyond Japan, where it was developed in the early part of the twentieth century by Mikao Usui during a period of fasting and meditation on Mount Kurama. It is used for treating all medical conditions, for improving quality of life or for preventing disease.

A treatment session would normally involve the fully clothed patient lying down on a massage table. Then the healer may or may not touch the client while transmitting healing energy. A session may last for about an hour and most patients would experience it as intensely relaxing.

There are several clinical trials of reiki and some of their results seem to suggest that this approach is beneficial for a range of conditions. However, most of this research is seriously flawed. For instance, many of these unreliable studies compare patients who elected to receive reiki with others who had no treatment at all. Any positive outcome in such a trial is likely to be due to a placebo effect, or to the attention those patients receive, and not necessarily to the reiki intervention itself. A critical analysis of the existing evidence therefore fails to demonstrate that reiki is effective.

There is, of course, a danger that reiki is used in serious conditions as a replacement for effective treatments, particularly as reiki practitioners claim to help any type of patient. There are, however, no direct risks associated with this approach.

Reiki is a popular form of spiritual healing, but it has no basis in science. The trial evidence fails to show its effectiveness for any condition.

Relaxation Therapies

Patients experience many alternative therapies as relaxing, e.g. meditation, hypnotherapy, autogenic training, massage, reflexology. While these therapies generate relaxation as a welcome side-effect, relaxation therapies are specifically designed to generate what is known as the 'relaxation response'. This term describes a pattern of reactions of the autonomic nervous system producing relaxation of the body and the mind. It is reflected in changes of physiological parameters, such as reductions of brain activity, heart rate, blood pressure, muscle tension, etc.

Relaxation techniques such as progressive muscle relaxation, visualization or imagery are practised by many alternative health practitioners, doctors, nurses, psychologists, psychotherapists or sports therapists. They are used to treat a range of conditions, including anxiety, stress, headaches, musculoskeletal pain, or they are employed to enhance physical or mental performance. The techniques are usually taught in supervised sessions; once the patient is able to elicit a relaxation response, he or she is advised to practise regularly at home. This, of course, requires time and dedication.

The evidence for relaxation therapies is mixed, and depends particularly on the condition under consideration. Relaxation treatments are effective for reducing stress and anxiety. Encouraging evidence also exists for treating insomnia, hypertension and menopausal symptoms. Whether relaxation treatments are helpful for controlling pain is still controversial, and they do not seem to be effective for chronic fatigue syndrome, irritable bowel syndrome, dyspepsia and epilepsy.

For patients with schizophrenia or severe depression, relaxation might aggravate their problem. Otherwise there seem to be no serious risks.

Relaxation techniques are helpful for a range of conditions. They are appreciated by many, not least because they put patients in charge of their own health.

Shiatsu

Shiatsu can be seen as the Japanese synthesis of acupuncture and massage. Literally it means finger (*shi*) pressure (*atsu*). It was founded by Tokujiro Namikoshi, who established the Japan Shiatsu College in Tokyo in 1940. At the age of seven, Namikoshi discovered the value of shiatsu when he treated his mother who suffered from rheumatoid arthritis.

The therapist typically uses his thumb to apply strong pressure on acupuncture points. Sometimes the palm of the hand or the elbow are also used. The treatment can be painful for the patient.

A shiatsu practitioner would start by making a diagnosis about the balance of the two life forces, yin and yang, so to some extent shiatsu is similar to traditional Chinese medicine. Depending on the findings, the practitioner would then apply pressure to points along yin or yang meridians. If the patient is diagnosed as having an excess of one, the therapist would tend to stimulate the other. By re-establishing balance, shiatsu practitioners believe they can treat many conditions.

As yin and yang, acupuncture points and meridians are not a reality, but merely the products of an ancient Chinese philosophy, shiatsu is an implausible medical intervention. However, like all massage techniques, it may generate relaxation and a sense of wellbeing.

There are virtually no clinical trials of shiatsu, but there is no reason to think that it is any more effective than a conventional massage. Shiatsu massage therefore seems to be a waste of effort and expense when compared to conventional massage.

Due to the high forces applied during treatment, injuries can occur. These range from bruises to bone fracture in the elderly with advanced osteoporosis. There are also reports of retinal and cerebral artery embolism associated with shiatsu massage applied to the neck or head.

Spiritual Healing

Many forms of spiritual healing exist: faith healing, intercessory prayer, reiki, therapeutic touch, psychic healing, Joheri healing, wart charming, etc. The common denominator is that healing 'energy' is channelled via the healer into the body of the patient. This 'energy' is supposed to enable the patient's body to heal itself. The term 'energy' needs to be put in inverted commas because it certainly is not energy as understood by scientists, but rather it has a spiritual or religious basis. All attempts to detect it have so far failed.

Healers view themselves as instruments of a higher power with healing ability bestowed upon them from above. Most state that they have no idea how their treatment works, but are nevertheless convinced that it does. The patient on the receiving end often feels warmth or tingling as the 'energy' apparently enters the body.

Consulting a healer usually involves a short conversation about the nature of the problem. The healer then starts the healing ritual. Initially this can be diagnostic by nature. For instance, the healer's hands may glide over the patient's body to identify problem areas. Eventually the healing starts, and 'energy' is supposed to flow. Many patients experience this as extremely relaxing, while healers often feel drained after a session. With other forms of spiritual healing, however, there is no personal contact between healer and patient. Sessions can be conducted at great distances, over the phone or the internet. Some healers offer their services for free, while others charge up to £100 for a half-hour session.

The concept of healing 'energy' is utterly implausible. Many clinical trials of various healing techniques are available. Some initially generated encouraging results, but about twenty of these studies are now suspected to be fraudulent. More recently, rigorous trials have emerged and shown that spiritual healing is associated with a large placebo effect – but with nothing more.

Spiritual healing is biologically implausible and its effects rely on a placebo response. At best it may offer comfort; at worst it can result in charlatans taking money from patients with serious conditions who require urgent conventional medicine.

Traditional Chinese Medicine

According to Traditional Chinese Medicine (TCM), all ill-health is viewed as an energy imbalance, while optimal health is a state of perfect balance, often symbolized by the yin–yang image. The aim of any therapy must be to restore the balance or to prevent any imbalance in the first place. For this purpose, TCM offers a range of treatments, including herbal mixtures, acupuncture, cupping, massage and diet, which are all discussed in more detail elsewhere in this book. All conditions are said to be treatable with TCM.

A TCM consultation will involve diagnostic techniques, such as tongue and pulse diagnoses. Although these techniques are also part of conventional medicine, TCM practitioners make unreasonably ambitious claims about their diagnostic power. Treatment will be tailored to the individual. One session would typically last 30–60 minutes, and treatment can be long-term, if not for life.

The TCM system is complex and not easy to evaluate. Thus its various elements are usually tested separately (see acupuncture in Chapter 2, for instance). Chinese herbal medicines usually contain a multitude of herbs which are individualized according to the specific needs of every patient. This approach has recently been tested in cancer patients and shown to be no better than placebo in alleviating symptoms. In another study, Chinese herbal medicine was tested in patients with irritable bowel syndrome against a standardized herbal prescription and against a placebo. The results suggested that individualized treatment is better than placebo in controlling symptoms, but not better than a standardized herbal medicine.

Some individual herbs used in TCM (e.g. liquorice, ginger, ginkgo) undoubtedly have pharmacological effects; some have even provided the blueprint for modern drugs. On the other hand, some Chinese herbal medicines are toxic (Aristolochia) and others may interact with prescription drugs. Chinese 'herbal' preparations may also contain non-herbal ingredients (e.g. endangered animal species), contaminants (e.g. heavy metals) or adulterants (e.g. steroids).

As you can see, TCM is difficult to evaluate. Some elements may be effective for some conditions, while other elements (e.g. cupping) are unlikely to offer any benefit above placebo.

Further Reading

The following books, articles and websites offer readers more information about the topics discussed in each chapter. Many of the references are books aimed at the general reader, but we have also included some key research papers, which can either be downloaded from the web or ordered at your local library. We have deliberately listed only a few of the main research papers relating to each alternative therapy, but these papers include references to many other pieces of research mentioned in this book. Additional references are available at www.trickortreatment.com

Chapter 1: How Do You Determine the Truth?

Wootton, David, *Bad Medicine: Doctors Doing Harm Since Hippocrates*, OUP, 2006.

Porter, Roy, *Blood and Guts: A Short History of Medicine*, Allen Lane, 2002.

Harvie, David, *Limeys: The Conquest of Scurvy*, Sutton, 2005.

Evans, I., Thornton, H., Chalmers, I., *Testing Treatments: Better Research for Better Healthcare*, British Library, 2006.

Doll, R., Hill, A. B., 'The mortality of doctors in relation to their smoking habits', *British Medical Journal* 1954; 228:1451–5.

Moore, A., McQuay, H., *Bandolier's Little Book of Making Sense of the Medical Evidence*, OUP, 2006.

Chapter 2: The Truth About Acupuncture

Kaptchuk, T. J., *The Web That Has No Weaver: Understanding Chinese Medicine*, McGraw-Hill, 2000.

Ernst, E., White, A., *Acupuncture: A Scientific Appraisal: A Scientific Approach*, Butterworth-Heinemann, 1999.

Evans, D., *Placebo: Mind Over Matter in Modern Medicine*, OUP, 2004.

Linde, K., et al., 'Acupuncture for Patients with Migraine: A Randomised Controlled Trial', *JAMA* 2005; 293:2118–25.

White, A., Rampes, H., Campbell, J. L., 'Acupuncture and related interventions for smoking cessation', *Cochrane Database of Systematic Reviews*, 2006.

Ernst, E., 'Acupuncture – a critical analysis', *J Intern Med* 2006; 259:125–37.

Chapter 3: The Truth About Homeopathy

Shelton, J. W., *Homeopathy: How It Really Works*, Prometheus, 2003.

Hempel, S., *The Medical Detective: John Snow, Cholera and the Mystery of the Broad Street Pump*, Granta, 2007.

Ernst, E., 'Evaluation of homeopathy in Nazi Germany', *Br Homeopath J* 1995; 84:229.

Maddox, J., Randi, J., Stewart, W. W., 'High-dilution experiments a delusion', *Nature* 1988; 334: 287–91.

Linde, K., 'Impact of Study Quality on Outcome in Placebo-Controlled Trials of Homeopathy', *Journal of Clinical Epidemiology* 1999; 52:631–636.

Shang, A., et al., 'Are the clinical effects of homeopathy placebo effects? Comparative study of placebo-controlled trials of homoeopathy and allopathy', *Lancet* 2005; 366:726–32.

Ernst, E., 'A systematic review of systematic reviews of homeopathy' *Br J Clin Pharmacol* 2002; 54:577–82.

Chapter 4: The Truth About Chiropractic Therapy

Salsburg, D., *The Lady Tasting Tea: How Statistics Revolutionized Science in the Twentieth Century*, Owl, 2002.

Ernst, E., Canter, P. H., 'A systematic review of systematic reviews of spinal manipulation', *J R Soc Med* 2006; 99:192–6.

Benedetti, P., MacPhail, W., *Spin Doctors: The Chiropractic Industry Under Examination*, Dundern, 2002.

Schmidt, K., Ernst, E., 'MMR vaccination advice over the Internet', *Vaccine* 2003; 21:1044–7.

Jonas, W. B., Ernst, E., 'Evaluating the safety of complementary and alternative products and practices', published in Jonas, W., Levin, J. (eds), *Essentials of Complementary and Alternative Medicine*, Lippincott, Williams & Wilkins, 1999.

Chapter 5: The Truth About Herbal Medicine

Hurley, Dan, *Natural Causes: Lies and Politics in America's Vitamin and Herbal Supplement Industry*, Broadway, 2006.

Fugh-Berman, A., *The 5-minute herbal and dietary supplement consult*, Lippincott, Williams & Wilkins, 2003.

Herr, S. M., Ernst, E., Young, V. S. L., *Herb-drug interaction handbook*, Church Street Books, 2002.

Ulbricht, C. E., Basch, E. M. (eds), *Natural Standard Herb & Supplement Reference: Evidence-Based Clinical Reviews*, Elsevier Mosby, 2005.

Whyte, J., *Bad Thoughts: A Guide to Clear Thinking*, Corvo, 2003.

Chapter 6: Does the Truth Matter?

Goldacre, B., *Bad Science*, Fourth Estate, 2008.

Ernst, E., Pittler, M. H., 'Celebrity-based medicine', *MJA* 2006; 185:680–81.

Colquhoun, D., 'Science degrees without the science', *Nature* 2007; 446:373–4.

Weeks, L., Verhoef, M., Scott, C., 'Presenting the alternative: cancer and complementary and alternative medicine in the Canadian print media', *Support Care Cancer* 2007; 15:931–8.

Appendix

Ernst, E., Pittler, M. H., Wider, B., Boddy, K., *The Desktop Guide to Complementary and Alternative Medicine: An Evidence-Based Approach* (2nd edition), Mosby, 2006.

Ernst, E., Pittler, M. H., Wider, B., Boddy, K., *Complementary Therapies for Pain Management: An Evidence-Based Approach*, Mosby, 2007.

Jonas, W. (ed.), *Mosby's Dictionary of Complementary and Alternative Medicine*, Mosby, 2005.

Hendler, S. S., Rorvik, D. (eds), *PDR for Nutritional Supplements*, Blackwell, 2001.

Useful Websites

The James Lind Library:
www.jameslindlibrary.org

The Cochrane Collaboration:
www.cochrane.org

Bandolier (evidence-based healthcare website):
www.jr2.ox.ac.uk/Bandolier

Focus on Alternative and Complementary Therapies (FACT)
www.medicinescomplete.com/journals/fact/current/

NIH, National Center for Complementary and Alternative Medicine
www.nccam.nih.gov

Healthwatch:
www.healthwatch-uk.org

Exeter University, Complementary Medicine Department:
www.pms.ac.uk/compmed/

Simon Singh's homepage:
www.simonsingh.net

Trick or Treatment? homepage:
www.trickortreatment.com

Acknowledgements

The conclusions presented in this book are based on decades of research conducted by thousands of medical researchers around the world. Without their efforts, it would be impossible to separate the effective from the bogus, and the safe from the hazardous.

We would like to offer particular thanks to the entire staff of the Department of Complementary Medicine, part of the Peninsula Medical School at the University of Exeter. They have supported this project from the outset and have always been generous with their advice and encouragement.

Despite the importance of the subject, there were times when it was unclear that this book would ever be published. We are indebted to our editors at Transworld and Norton, who had faith in our ambitions when others thought that alternative medicine was not a subject worthy of investigation. Sally Gaminara and Angela von der Lippe have been helpful and kind during a very intense eighteen months. Of course, we are also grateful to our literary agent Patrick Walsh, who is both a brilliant colleague and an excellent friend.

Last, and probably not least, our wives have both been remarkably wonderful, patient and lovely during the birth of this book. Anita and Danielle have shared our joys and our anxieties, our hopes and our fears. Thank you.

Picture Credits

James Lind © Wellcome Library, London

Florence Nightingale's polar chart © Wellcome Library, London

Model showing acupuncture needle entry points © Wellcome Library, London

Patient receiving acupuncture © Tek Image/Science Photo Library

Archie Cochrane © Cardiff University Library, Cochrane Archive, Llandough Hospital

Samuel Hahnemann © Science Photo Library

Oliver Wendell Holmes © Wellcome Library, London

John Snow's map of cholera deaths in Soho, 1854 © Royal Society of Medicine

Cervical spine © Sheila Terry/Science Photo Library

Daniel David Palmer © Science Photo Library

Field thistle © Wellcome Library, London

St John's wort © June Hill Redigo/Custom medical stock photo/Science Photo Library

Index